Computational Biology of Embryonic Stem Cells

Editor

Ming Zhan

The Methodist Hospital
Research Institute
USA

CONTENTS

FOREWORD

The stem cell field has seen several dramatic breakthroughs in the past decade. These include improvements in deriving ESC lines by a variety of methods as well as using adult cells and transforming them into pluripotent cells using subsets of transcription factors, microRNA and other regulatory proteins. These advance have resulted in an explosion in the number of pluripotent cells available.

Pluripotent stem cells have been derived from young and aged individuals, individuals of varying ethnic backgrounds, individuals carrying specific disease mutations as well as from cohorts of individuals carrying specific genetic traits.

It has quickly become clear that one needs to develop methods of comparing cell types and rapidly analyzing similarities and differences so that one can hone in on key regulatory pathways or molecules. Computational or mathematical simulation is also a powerful tool allowing quantitative description and systems exploration of pathways or networks. Moreover, methods or tools must be developed to dissect the large-scale and high-throughput "omics" data available for studies of these important cells.

It is in response to these needs that Dr. Ming Zhan has compiled this book on Computational Biology of Embryonic Stem Cells. Leading authorities in the field describe theories and techniques to extract and explore useful information from different cell populations that have been grown in different laboratories under differing conditions. The contributors also describe various algorithms and data mining techniques to understand the roles for microRNA, antisense, methylation and transcription factor regulation of stem cell proliferation and differentiation.

It is rare to find all of these methodologies and analyses well compiled in a useful and logically organized manner, and I hope that the readers of this book will find it as valuable as I did.

MahendraRao, Ph.D.

Director, Center for Regenerative Medicine
National Institute of Health
USA

PREFACE

Embryonic stem (ES) cells hold a great promise for regenerative medicine and treatment of illness such as neurodegenerative diseases, spinal cord injury, diabetes, and heart disease. In recent years, computational biology has significantly changed the landscape of the ES cell research, resulting in many significant discoveries. This book brings together reviews and reports from leading scientists to provide a comprehensive and updated introduction to the field of computational biology of ES cells.

The topics of this book is diverse, ranging from pure to applied computational biology research, and from integrated to systems biology studies on ES cells. We first introduce various bioinformatics algorithms and computational methodologies used for stem cell research. Banerjee *et al.* describe a method for reconstructing gene regulatory networks governing ES cell differentiation based ondiscrete temporal gene expression data.The method is formulated using an inherent feature of biological network, the sparsity of interconnection between transcription factors. Liu *et al.* review algorithms for structure learning of Bayesian networks and elucidating causal knowledge. The chapter answers the question about to what extent and by which means we can extract valuable biological knowledge from various experiment data. Newman *et al.* present AutoSOME, a novel unsupervised method for automatic clustering of large, high-dimensional data without prior knowledge of cluster number or structure. By applying this novel method on stem cell microarray data, the authors illustrate how to identify gene co-expression modules along with clusters of cellular phenotypes in a single step, and how to visualize transcriptome variation among stem cells using an intuitive network display. In introducing mathematical modeling studies on ES cells, Plopper *et al.* describe a multi-scale and multi-dimensional modeling of stem cells. The integrative modeling highlights how data gathered from one level can benefit research across multiple scales, addressing the challenge in analyzing increased amount of information of stem cells, which spans the entire breadth of biological fields, from molecular biology to population biology.

We next present focused computational analyses of the genome, transcriptome, proteome, epigenome and regulatory network of ES cells, and database for stem cell research. Chavez provides an overview of experimental techniques and computational methods for genome-wide methylation analysis, with focus on human ES cells. Sun & Zhan demonstrate transcriptional co-expression profiling of ES cells at global, pathway, and chromosome levels for exploring molecular mechanisms guiding ESC self-renewal and differentiation. Ji *et al.* describe computational analysis of alternative polyadenylation in ES cells and induced pluripotent stem (iPS) cells. The computational method described allows examining regulation of 3'UTR by alternative polyadenylation using DNA microarray data, and post-transcriptional regulation through cis-elements in 3'UTRs. Huelga & Yeo describe genome-wide detection of alternative splicing and highlight the importance of cis- and trans-factors in regulating alternative splicing in stem cells. Gunaratne & Tennakoon illustrate a microRNA-pluripotency gene network in ES Cells, and review the latest experimental technologies and computational algorithms forrevealing genomic and epigenetic changes associated with self-renewal and differentiation of ES cells. Lee *et al.*presentan integrated analysis of transcriptome and translation states for genome-wide identification of translationally regulated genes in ES cells. Sandie *et al.* demonstrate computational identification of non-coding antisense transcripts implicated in stem cell differentiation based on SAGE data and gene expression data. Han & Feng review the state-of-the-art ChIP-seq analysis tools developed for predicting ChIP-enriched genomic sites, and present a computational analysis of ChIP-seq data in ES cells. Finally, Hatzopoulos introduces the "Functional Genomics in Embryonic Stem Cells" (FunGenES) database. The database allows searching for gene expression and co-expression profiling data of mouse ES cells, as well as functional information of the relevant genes in embryonic development, adult homeostasis and disease.

The contributing authors of the book not only describe their researches and review the latest development of the field, but also discuss the future perspectives of the research. The book is a valuable reference and handbook for researchers and clinicians conducting stem cell research, and students and medical professionals interested in regenerative medicine, developmental biology, bioinformatics and computational biology.

Ming Zhan, Ph.D.

Associate Professor
Cornell University Weiss Medical College
Chief of Bioinformatics
The Methodist Hospital Research Institute

List of Contributors

Miguel A. Andrade-Navarro

Ottawa Hospital Research Institute, Ottawa, ON, Canada; Max Delbrück Center for Molecular Medicine, Robert-Rössle-Strasse, Berlin, Germany.

Ipsita Banerjee

Department of Chemical and Petroleum Engineering, University of Pittsburgh, Pittsburgh, PA, USA.

Pearl A. Campbell

Ottawa Hospital Research Institute, Ottawa, ON, Canada.

Lukas Chavez

Department of Vertebrate Genomics, Max-Planck-Institute for Molecular Genetics, Berlin, Germany.

James B. Cooper

Department of Molecular, Cellular and Developmental Biology, University of California, Santa Barbara, CA, USA.

Lin Feng

School of Computer Engineering, Nanyang Technological University, Singapore.

Preethi H. Gunaratne

Department of Biology & Biochemistry, University of Houston, Houston, TX, USA; Department of Pathology, Human Genome Sequencing Center, Baylor College of Medicine, Houston, TX, USA.

Jing-Dong J. Han

Chinese Academy of Sciences Key Laboratory of Molecular Developmental Biology, Center for Molecular Systems Biology, Institute of Genetics and Developmental Biology, Chinese Academy of Sciences, Beijing, China.

Xu Han

School of Computer Engineering, Nanyang Technological University, Singapore.

Antonis K. Hatzopoulos

Department of Medicine, Division of Cardiovascular Medicine, Vanderbilt University, Nashville, TN, USA.

Sebastian Hoersch

Informatics and Computing Core, Koch Institute for Integrative Cancer Research, Massachusetts Institute of Technology, Cambridge, MA, USA; Bioinformatics Group, Max Delbrück Center for Molecular Medicine, Robert-Rössle-Strasse, Berlin, Germany.

Mainul Hoque

Department of Biochemistry and Molecular Biology, Graduate School of Biomedical Sciences and New Jersey Medical School, University of Medicine and Dentistry of New Jersey, Newark, New Jersey, USA.

Stephanie C. Huelga

Bioinformatics Graduate Program, Stem Cell Initiative, Department of Cellular and Molecular Medicine, Institute for Genomic Medicine, University of California, San Diego, CA, USA.

Zhe Ji

Department of Biochemistry and Molecular Biology, Graduate School of Biomedical Sciences and New Jersey Medical School, University of Medicine and Dentistry of New Jersey, Newark, New Jersey, USA.

Winston Koh

Bioinformatics Institute, Agency for Science Technology and Research (A*STAR), Singapore.

Paul M. Krzyzanowski

Ottawa Hospital Research Institute, Ottawa, ON, Canada.

Melinda Larsen

Department of Biological Science, State University of New York, Albany NY, USA.

Qian Yi Lee

Bioinformatics Institute, Agency for Science Technology and Research (A*STAR), Singapore.

Yi Liu

Chinese Academy of Sciences Key Laboratory of Molecular Developmental Biology, Center for Molecular Systems Biology, Institute of Genetics and Developmental Biology, Chinese Academy of Sciences, Beijing, China.

Spandan Maiti

Department of Chemical and Petroleum Engineering, University of Pittsburgh, Pittsburgh, PA, USA.

Enrique M. Muro

Max Delbrück Center for Molecular Medicine, Robert-Rössle-Strasse, Berlin, Germany.

Aaron M. Newman

Department of Molecular, Cellular and Developmental Biology, University of California, Santa Barbara, CA, USA.

Gareth A. Palidwor

Ottawa Hospital Research Institute, Ottawa, ON, Canada.

Carolina Perez-Iratxeta

Ottawa Hospital Research Institute, Ottawa, ON, Canada.

George Plopper

Department of Biology, Rensselaer Polytechnic Institute, Troy, NY, USA.

Christopher J. Porter

Ottawa Hospital Research Institute, Ottawa, ON, Canada.

Feodor Price

Ottawa Hospital Research Institute, Ottawa, ON, Canada.

Michael A. Rudnicki

Ottawa Hospital Research Institute, Ottawa, ON, Canada.

PrabhaSampath

Institute of Medical Biology, Agency for Science Technology and Research (A*STAR), Singapore.

ReathaSandie

Ottawa Hospital Research Institute, Ottawa, ON, Canada.

Mandy Smith

Ottawa Hospital Research Institute, Ottawa, ON, Canada.

Yu Sun

Bioinformatics Unit, National Institute on Aging, NIH. Baltimore, MD, USA.

Vivek Tanavde

Bioinformatics Institute, Agency for Science Technology and Research (A*STAR), Singapore.

Keith Task

Department of Chemical and Petroleum Engineering, University of Pittsburgh, Pittsburgh, PA, USA.

Jayantha B. Tennakoon

Department of Biology & Biochemistry, University of Houston, Houston, TX, USA.

Bin Tian

Department of Biochemistry and Molecular Biology, Graduate School of Biomedical Sciences and New Jersey Medical School, University of Medicine and Dentistry of New Jersey, Newark, New Jersey, USA.

BülentYener

Department of Computer Science, Rensselaer Polytechnic Institute, Troy, NY, USA.

Gene W. Yeo

Bioinformatics Graduate Program, Stem Cell Initiative, Department of Cellular and Molecular Medicine, Institute for Genomic Medicine, University of California, San Diego, CA, USA.

Hong Yu

Chinese Academy of Sciences Key Laboratory of Molecular Developmental Biology, Center for Molecular Systems Biology, Institute of Genetics and Developmental Biology, Chinese Academy of Sciences, Beijing, China.

Ming Zhan

Bioinformatics Unit, National Institute on Aging, NIH. Baltimore, MD, USA (currently, The Methodist Hospital Research Institute, Houston, TX, USA).

CHAPTER 1

A Genetic Network Identification Algorithm Combining Experiment and Computation

Ipsita Banerjee[*], Keith Task and Spandan Maiti

Department of Chemical and Petroleum Engg, University of Pittsburgh, USA

Abstract: Embryonic stem cells (ESC) have potential to be used in future therapeutic applications due to their unlimited self-renewal capabilities coupled with their ability to differentiate into any cell type. Mathematical models of ESC have gained much attention in recent years for their ability to extract information and insight from this self-renewal and differentiation system which might be otherwise elusive when using experimental data alone. In this chapter, we first present a brief review of previous efforts to model the ESC system, including foci on single cells, populations, self-renewal and differentiation mechanisms, and signaling and gene regulatory networks (GRN). GRN identification in ESC is invaluable, as proper information on network connections can give insight on how stem cells differentiate and can help in the development of efficient differentiation to specific cellular phenotypes. Although there has been considerable work on network identification of bacteria and the ESC self-renewal circuitry, work is still limited on differentiation. We therefore present our work on reverse engineering the gene regulatory network in differentiating ESC. In our network identification algorithm, we incorporate the inherent biological feature of sparsity, the notion that a network favors as few connections as possible. Our algorithm consists of a bi-level formulation, in which the upper level predicts the network topology and minimizes the number of connections, while the bottom level estimates the kinetic parameters and minimizes the error between predicted and experimental profiles. We apply our bi-level formulation to the system of mouse ESC differentiating towards pancreatic lineage. The input to the algorithm was the expression dynamics of relevant transcription factors. We show that the predicted gene behavior is in very good agreement with the *in vitro* experimental data, and that many of the interactions in the reconstructed network and predicted effects of external perturbations have been reported in literature, even though this information was not used to train the model *a priori*. The predictive capability of the algorithm was further substantiated by modeling the effect of Foxa2 silencing on differentiation outcome and validating it experimentally by gene silencing experiments.

Keywords: Gene regulatory network, network identification, self-renewal, differentiation, embryonic stem cells, network sparsity, kinetic model, reverse engineering, network topology, transcription factors, perturbations.

*Address correspondence to Ipsita Banerjee: Department of Chemical and Petroleum Engg, University of Pittsburgh, 1242 Benedum Hall, 3700 O'Hara Street, Pittsburgh, PA 15261, USA; Tel: 412.624.2071; E-mail: ipb1@pitt.edu

1. INTRODUCTION

Embryonic stem cells (ESCs) are pluripotent cells which can give rise to any tissue type in the body. The primary characteristics of ESCs are that they can proliferate indefinitely and can differentiate to tissue specific lineages *in vitro*. The possibility of differentiation of ESCs to most of the tissue types has already been established, either by modulating the external cellular microenvironment or by genetic manipulation of the cells [1-3] Environment manipulations are typically achieved by addition of soluble factors to the culture media, modification of the substrate, or promoting cell-cell interaction by introduction of another cell or tissue type. However, what still remains a challenge to the community is the yield and homogeneity of the differentiated cell population and functionality of the differentiated cell-type. Another challenge lies in the lack of repeatability of the experiments, and even if some phenomena can be reproduced in several laboratories, the variability of the response is quite significant. Moreover, different cell lines are known to respond differently making it particularly challenging to establish a robust differentiation protocol. All these factors are limiting the full exploitation of the therapeutic potential of embryonic stem cells. Therapeutic use of stem cells would require a robust source of functionally differentiated cells, which requires an in-depth understanding of the mechanisms mediating the process of differentiation. Such understanding will be possible through a systematic approach integrating mathematical models with experimental data.

The past decade has seen notable contributions towards experimental and mathematical analysis of stem cell culture and differentiation process. Varied approaches have been used to interrogate different aspects of the system including cellular signaling network, gene regulatory network, single cell dynamics, population dynamics, self renewal *vs.* differentiation dynamics, *etc.* However, till date an integrated experimental and theoretical analysis of the embryonic stem cell system has been limited. Here we present a brief overview of such integrated analysis of the ESC system, followed by a detailed description of the methodology we have developed in identifying the regulatory network governing the process of stem cell differentiation.

Viswanathan *et al.* [4] proposed a single cell model of the ESC system which would account for the heterogeneity in the cell population. They based their model on number of ligands/receptors per cell, and predicted the behavior of ESC self-renewal and differentiation, and the system's response to different exogenous stimuli. Such analysis has potential use in selecting specific tuning parameters while guiding ESCs towards a specific fate [4, 5]. Prudhomme *et al*. [6] developed an ordinary differential equation based kinetic model to quantify the differentiation dynamics in response to combinations of different extracellular stimuli. Based on experimental data of ESC response to different combinations of extracellular matrix and cytokines, the authors estimated kinetic rate constants for each culture conditions. Such kinetic studies can be used to elucidate the mechanism of differentiation in response to specific cues. In an attempt to understand the signaling network in place during the process of differentiation, the same group [7] performed a thorough systematic analysis of how the intracellular signaling relates to different extracellular cues during differentiation of mouse ESCs. A partial-least-squared multivariate model was built to show the role of signaling proteins in self-renewal, differentiation, and proliferation of stem cells. In a follow up work, Woolf *et al.,* [8] investigated the signaling network to determine the "cue-signal-response" interactions through a Bayesian network algorithm. The nodes of the network are assigned to be an extracellular stimulus, a signaling protein, or a cell response, following which the model identified interconnections between nodes without being explicit about the nature of connections (inhibition, induction). Another example of an integrated experimental and mathematical analysis of ESCs is the work of Willerth and Sakiyama-Elbert in quantifying the effect of neurotrophin-3 (NT-3) on the differentiation of ESC-derived neural progenitor cells to neurons [9]. These results provide a detailed analysis of NT-3 threshold levels required to induce MAPK cascade activation and subsequent neuronal differentiation.

While there have been important contributions towards integrated analysis of signaling pathways in ESC differentiation, parallel efforts in analyzing the gene regulatory networks are largely confined to theoretical analysis only. Most widely studied is the Oct4-Sox2-Nanog network which regulates the self-renewal and differentiation events of ESCs. Notable studies of this network include the study by Chickarmane *et al*. [10], which reports identification of a bistable switch in the Oct4-Sox2-Nanog network leading to a binary decision of the cells to self-renew or differentiate. In a follow up work [11].the authors further extend the model to incorporate lineage specific differentiation namely to endoderm and trophectoderm. MacArthur *et al.* [12] also analyzed the Oct4-Sox2-Nanog network coupled with a lineage specification network to investigate the induction of pluripotent cells from somatic cells.

Our group has developed an integrated approach in identifying the transcription factor network governing differentiation from experimental data of the ESC differentiation system, details of which will be presented in this chapter. While rare in ESC systems, network identification problem has seen significant success in analyzing bacterial and yeastnetworks [13, 14] However, the generalization of these methods to the inference of networks in higher eukaryotes is not always obvious. Typically, network identification problem is mathematically ill-posed; certain properties of the network under investigation need to be incorporated in the formulation for the uniqueness of the solution. We have developed our algorithm of network identification primarily by exploiting the notion of network sparsity, known to be common in many biological systems. Our method identifies both the cell-intrinsic regulatory architecture along with the effect of external perturbation to this network, both of which are of significant importance in the process of stem cell differentiation. As will be discussed subsequently, the approach will be beneficial for the development of targeted experimental protocols for the production of cells with a pre-specified fate.

2. NETWORK IDENTIFICATION FOR DIFFERENTIATING ESCs

Akin to *in vivo* development the process of ESC differentiation proceeds in specific and distinct stages. It is known that developmental regulatory network is typically organized in a distinctive cascade of control [15] that enables the subdivision of the entire complex network into a number of smaller subsets or modules. Each module is under the control of a signature gene or 'hub' that plays a central role in directing the cellular response to a given stimulus. This observation reinforces the discretization of the entire process of stem cell differentiation into specific stages, and investigation of one such stage in isolation. Furthermore, the mathematical formulation for network identification is developed based on the rationale that network sparsity characterizes the regulatory architecture governing development. Network sparsity has been experimentally observed in the visual system of primates [16], auditory system of rats [17], and olfactory system of insects [18]. The notion of network sparsity is envisaged here as the governing criterion determining the regulatory network of differentiating embryonic stem cells, and a formal mathematical structure has been developed to analyze such systems. The network identification problem is applied to a system of embryonic stem cells differentiating towards pancreatic lineage following an integrated experimental and mathematical approach as illustrated in Fig. (**1**). Briefly, the mathematical model is developed based on experimental data of ESC differentiation, following which the model is used to predict a possible pathway for further differentiation which is validated by subsequent experimentation.

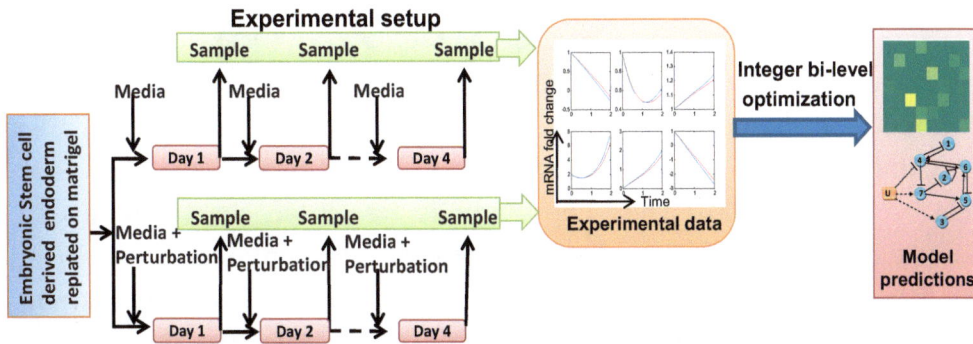

Figure 1: Schematic representation of the integrated experimental and computational approach applied to understand the regulatory interaction governing stem cell differentiation. The differentiating cell population was sampled every day and analyzed for 13 transcription factors. A bi-level integer programming formulation is solved to identify the regulatory interactions that accurately reproduce the experimentally observed transcription factor dynamics.

2.1. Mathematical Model

Differentiation of embryonic stem cells is typically induced by manipulating the cellular external environment *via* substrates, growth factors, chemical inducers/ repressors, *etc.* Such external perturbations affect the intrinsic cellular state resulting in a lineage specific differentiation. Hence what will be of interest is (i) an understanding of the intrinsic regulatory network of the ESCs and (ii) the effect of external perturbations on the network. First step in analyzing the regulatory network is a mathematical representation of the network, which can be done in a variety of ways including Boolean logic [19], Bayesian networks [20], graph theory [21], and ordinary differential equations [22]. We have modeled the gene expression profile as a time continuous dynamical system by representing it as a system of coupled ordinary differential equations [23, 24]:

$$\dot{\mathbf{X}} = f(\mathbf{X}) \tag{1}$$

where $\mathbf{X} = \{x_1, x_2, \ldots, x_n\}$ represents the array of n mRNA concentrations of interest, and dot denotes the differentiation with respect to time. In the present study, $f(\mathbf{X})$ is modeled by a linear set of equations given by:

$$\dot{\mathbf{X}} = \mathbf{AX} + \mathbf{BU} \tag{2}$$

where $\mathbf{A}$ is the $n \times n$ connectivity matrix, $\mathbf{B}$ is the $n \times p$ matrix representing the effect of p perturbations on n genes, and $\mathbf{U}$ is a $p \times 1$ vector representing p perturbations. The connectivity matrix represents which of the genes are interconnected in the network and also the strength of connection. Matrix $\mathbf{A}$ is intrinsic to the cell, while vector $\mathbf{B}$ represents the environmental influence. In the context of stem cell differentiation, u_i will represent the concentration of ith growth factors/inducers/inhibitors used in the differentiation process, and b_{ij} will reflect the effect of these factors on j^{th} gene in the network. In order to discretize Equation 2 in time, in a manner such that experimental observations can be incorporated, we chose bi-linear transformation because of its stability and low computational cost [25]. Using this transformation, Equation 2 is converted to its discrete form:

$$\mathbf{X}(t_{k+1}) = \frac{2 + \mathbf{A}\Delta t}{2 - \mathbf{A}\Delta t} \mathbf{X}(t_k) + \frac{2\mathbf{B}\Delta t}{2 - A\Delta t} \mathbf{U}(t_k) \tag{3}$$

$$= \mathbf{A}_d \mathbf{X}(t_k) + \mathbf{B}_d \mathbf{U}(t_k)$$

where subscript k denotes the value of a quantity at the current sampling point, and $k+1$ is the next sampling point. In the above formulation, the input parameters are the experimentally determined values of gene expression levels at different experimental time points $\mathbf{X}(t_k)$ as well as the external perturbation $U(t_k)$. The unknown parameters to be determined are the connectivity matrix $\mathbf{A}$ and the effect of the external perturbation, $\mathbf{B}$. Given enough sampled observations of $\mathbf{X}(t_k)$, the estimation problem becomes well posed and a solution providing the best fit in the least square sense can be computed by minimizing the following objective function:

$$\min_{\mathbf{A},\mathbf{B}} \left\| \mathbf{X}(t_k+1) - \mathbf{A}_d\mathbf{X}(t_k) - \mathbf{B}_d\mathbf{U}(t_k) \right\|^2 \tag{4}$$

However, this formulation becomes ill-posed in the absence of sufficient experimental data points, which is commonplace in biological systems, specifically in stem cell differentiation. Typically in the literature, at this stage, a dimension reduction technique such as Principal Component Analysis (PCA) or Singular Value Decomposition (SVD) is employed [23] to map the original problem space to a lower dimensional subspace. Such techniques, although mathematically tractable, will invariably lead to a loss of information resulting from dimension reduction.

We have developed an alternate strategy to address such underdetermined systems based on inherent properties of biological networks: that of *network sparsity*. Among the n^2 possible regulations between n nodes of a network, nature would select a robust network with fast response time that maximizes compartmentalization of cellular processes, hence giving rise to a *small world* network. Using the hypothesis of network sparsity, the reverse engineering problem is formulated as a bi-level mixed integer programming problem [26], where the upper level determines the network topology by promoting network sparsity using an integer programming formulation while the lower level optimizes the strength of the existing connections as determined by the upper level. Thus, at the end of the procedure we will arrive at a detailed description of the network, consisting of both the topology and strength of interaction governing the dynamics of the gene expression profile. The problem formulation is given by:

$$\min \sum_{i,j=1}^{n} \lambda_{ij}$$

subject to : $\qquad\qquad\qquad\qquad\qquad\qquad$ (5)

$$\arg\min_{\mathbf{A},\mathbf{B}} \left\| \mathbf{X}(t_k+1) - \Lambda\mathbf{A}_d\mathbf{X}(t_k) - \mathbf{B}_d\mathbf{U}(t_k) \right\|^2 \leq tolerance, k=1,\ldots,m$$

where Λ represents $n \times n$ binary variables corresponding to each component of the connectivity matrix $\mathbf{A}$. $\lambda_{ij} = 0$ implies no influence of gene j on gene i, while $\lambda_{ij} = 1$ implies that gene j influences gene i. The objective of the upper level formulation is to minimize the total number of network connections, given by $\sum \lambda$, which essentially reduces the density of the connectivity matrix. The upper level thus constitutes an L_0 norm minimization problem, where the number of elements of the matrix $\mathbf{A}$ is minimized to promote sparsity. The constraint evaluated at the lower level is also a minimization problem which ensures that the predicted profile matches the experimental data within user defined accuracy specified by a tolerance. The optimization variables of the lower level are continuous, and it determines the strength and the nature (inducer/ inhibitor) of the connectivity matrix.

For a network of n genes, the connectivity matrix $\mathbf{A}$ consists of n^2 elements, and the vector $\mathbf{B}$ consists of n elements. Hence the upper level contains n^2 binary variables and is solved using a combinatorial optimization technique, namely Genetic Algorithm (GA). This route is chosen since L_0 minimization is an NP hard problem which is better suited to be solved using combinatorial approaches rather than approximation algorithms [27]. Although there is no efficient algorithm for hard combinatorial problems, Evolutionary Algorithms have been found to be efficient in finding an approximate solution [28]. Implementation of GA typically requires coding the continuous variables as bits of binary strings, and decoding the binary bits back to continuous form. However, the present integer programming formulation is particularly conducive to Genetic Algorithm since the binary optimization variables could be directly encrypted in the representative chromosome of the algorithm, hence avoiding additional steps of coding and decoding of continuous variables to binary format. Number of variables to be optimized in the lower level is determined at the upper level as:

$\sum\limits_{i,j=1}^{n} \lambda_{ij}$. Upper level integer programming essentially reduces the number of

variables to be optimized in the lower level, and as a consequence the estimation problem remains well posed even with a small number of experimental observations. An additional constraint is imposed on the least square minimization program to ensure that the number of estimated parameters does not exceed experimental data points:

$$1 \leq \sum_{i,j=1}^{n} \lambda_{ij} < n \times (p+1) \times (m-1) \qquad (6)$$

where m is the number of time points, and n and p are as defined before. The above constraint implies that the number of variables to be optimized in the lower level should be less than available data thus ensuring that the problem is solvable. Hence the available experimental data dictates the size and density of the connectivity matrix that can be evaluated. In the absence of external perturbation, the connectivity matrix can be determined with sparsity $< \dfrac{m}{n}$. To determine a fully connected matrix the required time points are: $m \geq n$.

2.2. Experimental Method

The above formulation is applied to a system of embryonic stem cells differentiating towards the pancreatic lineage. Pancreatic organogenesis, as with any developmental system, occurs in linear cascade of distinct stages starting with endoderm commitment, followed by pancreatic progenitors, endocrine progenitors, and finally to mature endocrine cells. In the *in vitro* differentiation of ES cells, a similar sequence is reproduced. The mathematical analysis of regulatory network is similarly treated as a cascade of events, of which we concentrate on a single stage of differentiation, that of pancreatic progenitor commitment, and analyze it for the relevant gene expressions.

The mouse ESCs are differentiated to endoderm-like cells by co-culturing them with primary hepatocytes [29]. The endodermal cells were harvested from the co-culture and replated on matrigel to induce pancreatic lineage as verified by Pdx-1 expression. Early pancreatic differentiation has been reported to be inhibited by Sonic Hedgehog (*Shh*) signaling [30] which in turn can be inhibited by Cyclopamine, a known repressor of *Shh*. The differentiating population was treated with Cyclopamine which acts as the external perturbation described in the section above. Data from two parallel experiments were considered: (a) the control case cultured in a differentiation media and (b) Sonic Hedgehog (*Shh*) inhibition by supplementing differentiation media with Cyclopamine (external perturbation). Both these conditions were analyzed for transcription factors reported to be relevant for early pancreatic differentiation. A thorough analysis of the literature for early pancreatic markers ([31] and references therein) reveals 13 major transcription factors (TF) that primarily constitute the TF network at the pancreatic progenitor stage. Hence our population of ESC derived pancreatic progenitor cells were also analyzed for these 13 TFs. Fig. (**2**) illustrates the fold change in mRNA levels of these transcription factors over the differentiation time, both for the control and the perturbed conditions. Each transcription factor is represented as fold change of expression levels compared to day 1 of differentiation. The normalized mRNA expression dynamics serves as the input to the network identification algorithm.

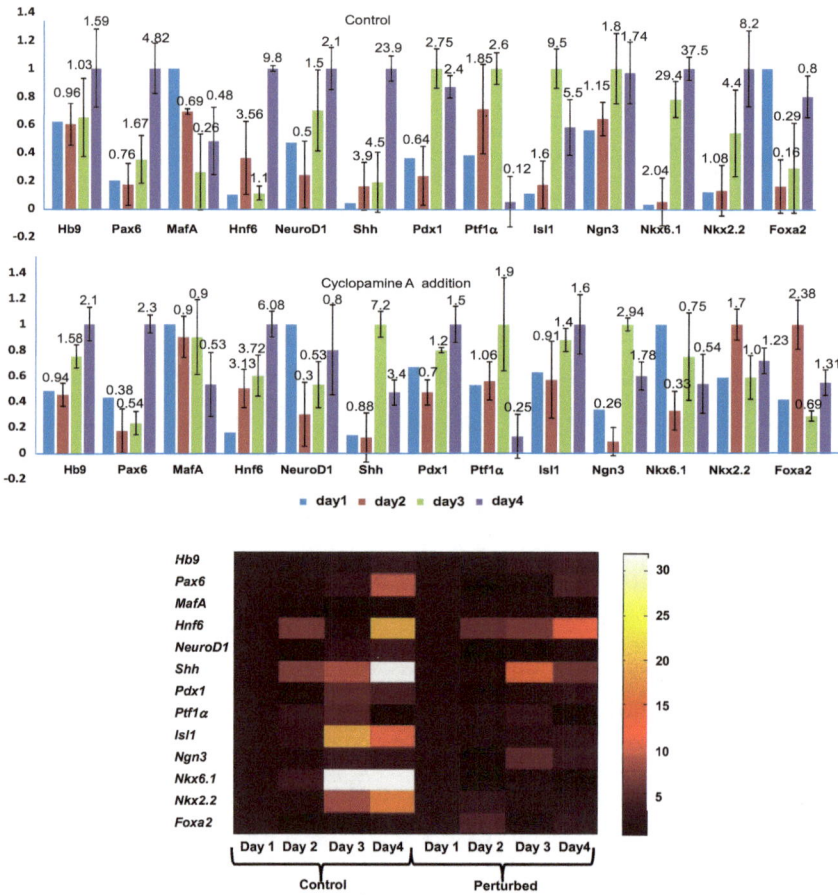

Figure 2: Experimental data for ES cells differentiating to the pancreatic lineage, cultured in (i) DMEM, 10% FBS and (ii) DMEM, 10% FBS, KAAD Cyclopamine A (external perturbation). Cells were harvested each day and analyzed for 13 relevant TFs. The data represents the relative fold change in TFs as obtained by qRT-PCR and reported as $2^{-\Delta\Delta ct}$ values. (Top) bar graph represents the scaled (0-1) value of relative fold change and the value above each bar represents the actual fold change. (Bottom) colormap representation of the TF dynamics.

3. RESULTS

3.1. Identification of Transcription Regulatory Network of ES Differentiation

The bi-level optimization problem is solved with mean value of the experimental data as the input to identify: (i) 13×13 matrix **A**, the regulatory interaction between measured transcription factors including both network topology and connectivity strength, and (ii) 13×1 vector **B**, the effect of external perturbation (Cyclopamine supplementation) on the regulatory network.

The upper level topology optimization problem leads to an integer programming problem with 169 binary variables representing 13x13 network connectivity, which is solved using Genetic Algorithm. The efficiency of the algorithm depends on appropriate choice of starting population, as well as other involved parameters. The initial population size plays an important role in quality and efficiency of the algorithm. A small population size may lead to local convergence or extremely large number of generations. To avoid these problems, a population size of 20 was chosen, and the algorithm was evolved over 200 generations using a tolerance value of 1.0. The tolerance value dictates how closely the predicted profiles are required to match the experimental data. The crossover probability was chosen to be at a standard value of 0.5, and the chosen mutation probability of 0.02 was expected to maintain diversity in population. For each combination of binary variables specifying the network topology in the upper level, the lower level regression problem is solved to optimize the connectivity strength against experimental data. Effect of the external perturbation is considered in the lower level as continuous variables, giving rise to a total of $13 + \sum \lambda_i$ continuous variables. Dynamics of the gene expression predicted by the reconstructed network **A** is illustrated in Fig. (**3**). Observe from this figure that the computationally predicted gene expressions show an excellent agreement with the experimental data. However, agreement of the mRNA profiles is imposed as a constraint to the lower level optimization problem, which depends on the chosen value to tolerance. Thus, although necessary, it cannot be judged as a sufficient condition indicating the accuracy of reconstruction.

Optimal reconstructed network obtained by solving the upper level integer programming problem results in 54 out of 169 connections, amounting to 68 %

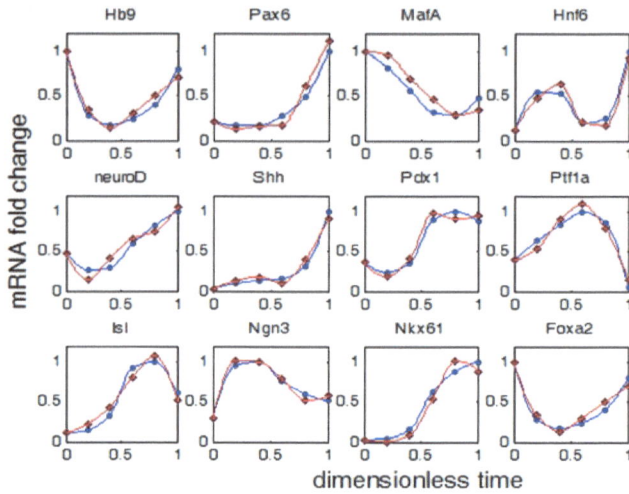

Figure 3: Comparison of mRNA profiles predicted by the reconstructed network (blue) with the experimental data (red). Network is reconstructed using the proposed bi-level integer programming formulation. An error tolerance of 1.0 captures the experimental data points with excellent accuracy.

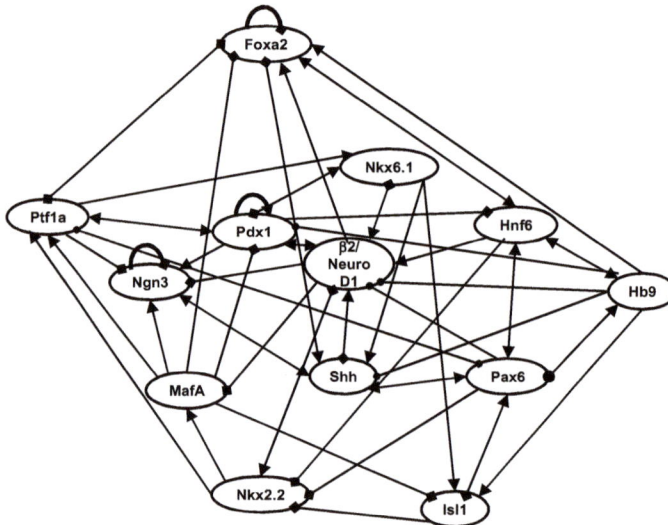

Figure 4: The reconstructed network **A**, obtained by solving the bi-level mixed integer programming problem, maximizing sparsity at the upper level and minimizing least square error at the second level. The arrow ⟶ indicates induction and the square ⟶• represents repression. A •⟶ B indicates that B is induced by A while A is repressed by B.

sparsity as represented in Fig. (**4**).The normalized strength of connectivity for each pair-wise network connection is depicted in Fig. (**5**). A broad range in connectivity strength, varying from 0.1 to 40, is observed in the mathematically constructed network. Accordingly, the pairwise connections are categorized in 3 groups depending on their connectivity strength, as depicted in Fig. (**5**). Only 9 of the 54 connections exhibited high connectivity strength in the range of 10 – 40 normalized values; 18 connections had a medium strength in the range of 5 – 10 while the rest of the 27 connections had a value lower than 5. Among the weaker connections, only the ones with values higher than 1 are shown in Fig. (**5**). In order to evaluate the sensitivity of the optimal reconstructed network to the experimental noise, a sensitivity analysis was performed by perturbing each of the experimental data points by 10% and evaluating the corresponding perturbation in the network connectivity. The bars in Fig. (**5**) represent the overall sensitivity of each of the pair-wise connectivities to all the experimental perturbations. It is observed that the optimal network is quite robust against experimental noise, since the sensitivity of most of the network connections remain bounded within 10% of the nominal value of the imposed perturbation.

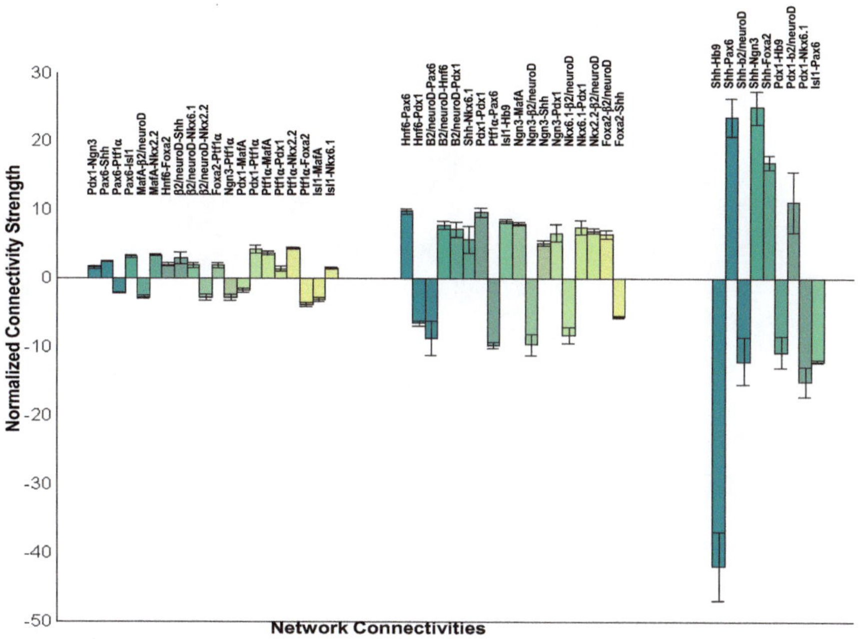

Figure 5: Normalized connectivity strength of the optimal network reconstructed by solving the bilevel optimization problem. Connectivities with insignificant strength have been omitted from the figure. Pairwise connectivities have been clustered into 3 groups depending on the strength of the connectivity for ease of viewing. Bars represent the sensitivity of a specific connectivity to experimental noise.

3.2. Effect of Environmental Perturbation on Gene Network

The objective of the above analysis was to determine the intrinsic regulatory network governing the process of differentiation, along with the influence external inductions have on the network. Such effects of Cyclopamine on the 13 transcription factors, given by vector **B**, are illustrated in Fig. (**6**). The strength of influence is depicted by the thickness of connecting lines, strength being directly proportional to the thickness. It is worth noting that out of the 13 elements of **B**, only 3 were of appreciable magnitude and the others were negligible. Overall, the strongest effect of Cyclopamine is predicted to be in the inhibition of *Shh*. Inhibition of *Ngn3* and up-regulation of *Nkx2.2*are also predicted, but with much lower strength.

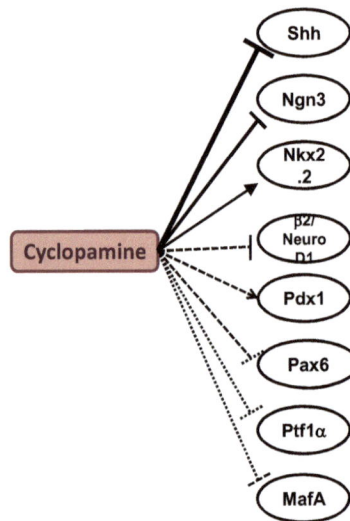

Figure 6: The predicted effect of external perturbation (Cyclopamine) on the gene regulatory network. The thickness of the arrow depicts the strength of connection. The network reconstruction algorithm could accurately predict the most prominent effect of Cyclopamine in inhibiting sonic hedgehog gene.

In vivo studies have reported that Cyclopamine addition leads to ectopic pancreas development [30], resulting from the inhibition of Hedgehog signaling. Both *in vivo* and *in vitro* studies have confirmed that the mechanism of *Shh* inhibition by Cyclopamine is indirect, resulting from blocking smoothened (*Smo*) function [32, 33]. Although such inhibitory effect of Cyclopamine on *Shh* is an established phenomenon, we did not provide this information *a priori* to the simulation in order to verify the predictive capability of our algorithm. As illustrated in Fig. (**6**), this algorithm could successfully identify the effect of Cyclopamine addition as being inhibition of Sonic Hedgehog. It is important to note that the effect of Cyclopamine on *Shh* is not direct, but through an indirect signaling cascade. Our model was not provided with enough details to capture the entire signaling pathway of Cyclopamine, but even then our formulation could accurately capture the resultant response of Cyclopamine on *Shh*. This is extremely crucial, since the effect of environmental perturbation on the differentiating cells is likely to be indirect. However, our primary interest is the altered functional behavior of the system in response to these perturbations which our model could predict accurately.

3.3. Identified Network Captures Known Interactions

The network which was identified using the bi-level programming approach revealed significant agreement with existing literature reports. The reconstructed network is analyzed primarily with respect to the pancreatic duodenal homeobox gene-1 *(Pdx-1)*, a master regulator for both pancreatic development and maturation to β-cell phenotype [31]. Fig. (7) illustrates the key comparison of the predicted connections with experimentally observed connections reported in literature. The strength of the predicted connections is depicted by the thickness of the connecting arrows. Overall, an excellent agreement between our reconstructed network and literature reports can be readily observed.

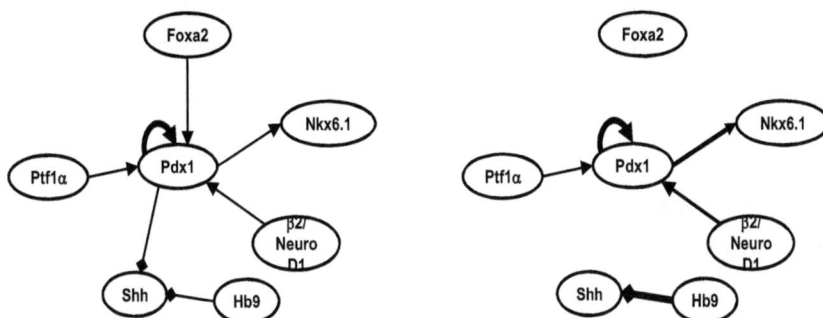

Figure 7: Comparison of the regulatory interactions of *Pdx-1* as reported in literature (left) with the reconstructed network (right). The strength of interactions in the predicted network is depicted by the thickness of the arrow. The reconstructed network could adequately capture many of the known interactions, without *a priori* information.

During development, islet progenitors arise from $Ptf1\alpha$-$p48^{+}/Pdx1^{+}$ cells and $Ptf1\alpha$ has been reported to directly bind to *Pdx1* promoter [34]. Our differentiation scheme results in strong up-regulation of *Ptf1α* and the network analysis reveals the positive induction of *Pdx1* expression by *Ptf1α*.

Pdx1 has also been reported to be directly activated by the transcription factors *NeuroD1* and *Foxa2* [35]. The positive induction of *Pdx1* by *NeuroD1* is accurately captured by the model, while the effect of *Foxa2* is more indirect. In addition, the strong autoregulatory effect of *Pdx1* in inducing its own transcription is an established phenomenon, which is accurately captured by the model prediction.

Targeted disruption of the *Pdx1* gene in β cells leads to reduced activities of *Pdx1* regulated genes such as *Nkx6.1* and *Glut2* [36]. A similar phenomenon of *Pdx1* induction of the *Nkx6-1* gene is also observed in the model prediction. As discussed before, one of the earliest events in pancreatic organogenesis is the repression of Sonic Hedgehog (*Shh*) by the notochord, which in turn promotes *Pdx1* expression in adjacent pancreatic endoderm [37]. In parallel, the combined action of *Pdx1* and *Hb9* inhibits *Shh*expression. The model captures a similar effect of *Shh* inhibition by *Hb9*, over and above the suppression of *Shh* by Cyclopamine.

Endocrine cells originate from lineage-committed progenitors marked by the transcription factor neurogenin 3 (*Ngn3*) [38], which has been shown to negatively regulate its own promoter [39]. The negative auto-regulatory effect of *Ngn3* is correctly predicted in the simulated network.

Nkx2.2 also drives endocrine differentiation and is controlled by alternative promoters at different cellular stages [40]. During progenitor and endocrine cell stages, *Ngn3* and *NeuroD1* have been shown to activate *Nkx2.2* respectively. The present model identifies *NeuroD1* as the inducer of *Nkx2.2* but not *Ngn3*, suggesting the differentiation stage being endocrine cellular state. The inactivation of *Nkx2.2* gives rise to endocrine-like cells lacking Insulin or *Glut2*, but expressing other endocrine markers such as Amylin and *Isl1*. The current model also indicates that *Isl1* may have some effect in negative regulation of *Nkx2.2*. White *et al.*, 2008, attempted to identify the regulatory structure of pancreas development and reported the positive effect of β*2/NeudoD1* in the up-regulation of *Nkx2.2* and *Foxa2*. *NeuroD1* was also shown to bind to *MafA*. All of these effects of *NeuroD1* are captured in the present reconstructed network.

These results indicate that the proposed algorithm developed on the notion of sparsity of biological networks can successfully extract from the experimental data many of the known interactions which have been independently reported in literature. It is important to note here that the reconstruction algorithm relies on the subset of the transcription factors used in the input data points. Thus the quality and resolution of the reconstructed network will largely depend on the input data provided to the model. However, even in the absence of precise details of intermediate steps, the model could adequately capture the overall behavior of the system, as demonstrated with the case of *Shh* inhibition.

3.4. Network Prediction and Experimental Validation

While the reconstructed network agrees well with literature reports on pancreatic developmental networks, the full potential of the model can only be exploited in its predictive capability. In order to test the predictive capacity of the derived model, we used the model to determine the subsequent differentiation pathway in the pancreatic lineage. The current network is determined for the pancreatic progenitor stage controlled by *Pdx1*, which is followed by the endocrine progenitor stage controlled primarily by *Ngn3*. We used the reconstructed model to identify a pathway which will significantly up-regulate *Ngn3* expression, thereby inducing endocrine differentiation.

This prediction is achieved by solving Equation (2) with mathematically derived **A** and **B** and a proper choice of **U**. The objective was to identify a gene which needs to be down-regulated for maximal up-regulation of *Ngn3*. The effect of silencing the ith gene will be predicted by adjusting u_j , components of **U** as

$$u_j = 0\Big|_{j=1,n;\, j\neq i}\,;\; u_i = -S$$

where $-S$ represents appropriate down-regulation of the ith gene. Each of the 13 genes was downregulated and the corresponding up-regulation of *Ngn3* was recorded. This exercise identifies down-regulation of *Foxa2* to be a likely mechanism in up-regulation of *Ngn3* expression levels. In the absence of literature reports relating such an interaction, the validity of this prediction is verified by performing concurrent experiments by silencing *Foxa2* gene in the differentiating stem cell population. Fig. (**8a**) represents the dynamic response of the system to *Foxa2* down-regulation and compares the model predictions with experimental observations. Fig. (**8b**) illustrates the colorbar representation of the comparison, the predicted *versus* actual effect of *Foxa2* silencing on the population of differentiating ESCs. Fig. (**8b**) clearly shows that the most significant effect of *Foxa2* silencing is the up-regulation of *Ngn3* (~10 folds). This observation compares extremely well with the model prediction of 8 folds up-regulation. *Foxa2* silencing also resulted in significant down-regulation of *MafA* genes, which is also correctly predicted by our reconstructed network, although the magnitude of down-regulation is somewhat underpredicted.

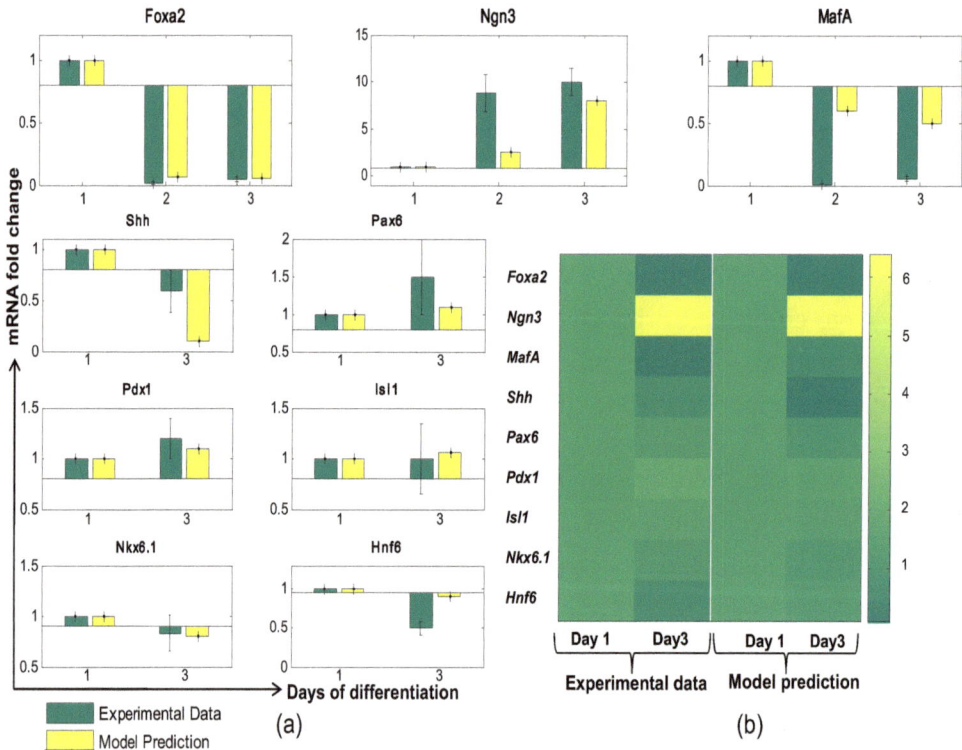

Figure 8: Effect of silencing of Foxa2 gene on the other components of the regulatory network. (a) Comparison of the model predicted effects of Foxa2 down-regulation with experimental data of Foxa2 silencing. The time points analyzed are: 24 hrs, 36 hrs and 52 hrs after initial transfection, depicted as 1, 2 and 3 in the x-axis respectively. (b) Colormap comparison of fold changes in gene expression levels between reconstructed model and experimental observation.

Silencing of *Foxa2* had less dramatic effect on many of the other genes, although most of them were affected to some extent. Quite encouragingly, our model could accurately predict most of these effects in both nature and strength. For example, both *Pax6* and *Pdx1* were up-regulated by *Foxa2* silencing, as also predicted by the network. *Shh* and *Hnf6* were both down-regulated by *Foxa2* silencing, which is correctly predicted by the model in nature but not in magnitude. The model overpredicts the down-regulation of *Shh*, while underpredicts the down-regulation of *Hnf6*. Both *Isl1* and *Nkx6.1* were quite insignificantly affected by the *Foxa2* silencing, which again is accurately predicted by the model as well.

Overall, the mixed integer bi-level programming approach based on the notion of sparsity succeeds in accurately capturing the most important and significant connections in the system, although prediction of the weaker connections can be less accurate. The proposed method can determine sufficiently accurate networks with very restricted experimental data, a feature which makes it extremely attractive to stem cell based applications.

4. DISCUSSION

A bi-level mixed integer programming formulation for reverse engineering gene regulatory network is presented in this chapter. The developed framework can accurately capture the regulatory interactions solely from the data of gene expression profiles, without any *a priori* information regarding regulatory interactions between transcription factors. The network architecture derived by promoting sparsity could adequately predict experimentally observed regulations reported in the literature. It could also accurately predict the perturbation required to induce subsequent differentiation, an outcome confirmed by concurrent experiments.

The gene regulatory network determined by the proposed bi-level methodology could reliably capture the known effects of *Shh* inhibition of Cyclopamine, along with key governing features of pancreatic organogenesis. The predicted network derived on the notion of sparsity had excellent predictive capability even outside the domain of experimental data used to determine the network. The presented method offers significant advantages in analyzing a differentiating cell population, where the cells are typically induced to a specific lineage by exposing them to different environmental conditions with respect to extracellular matrix, growth factors, and chemical inducers or repressors. The method can efficiently utilize this information in analyzing the governing network and does not rely on more laborious gene knockout data. This is even more pertinent in stem cell studies since knockout of certain key genes can have severe consequences with respect to cell survival and differentiation. However, in the event of availability of such data it can be easily incorporated into the network identification formulation. Moreover, the bi-level formulation ensures efficient utilization of experimental data by essentially reducing the number of variables being optimized in the inner loop.

Although developed and illustrated in the context of differentiating population of ESCs to pancreatic lineage, the mathematical framework is general enough to be applicable to any system of stem cells, embryonic or adult, to any lineage. Although there is considerable information regarding specific roles of transcription factors at different stages of organogenesis, very little information is currently available on how transcriptional networks are organized within these cells. The methodology will be instrumental in identifying such transcriptional networks and the environmental effect on such networks, which will have potential application in designing *in silico* protocols for stem cell differentiation. This approach will be particularly useful in identifying regulatory networks in data-limited systems like stem cell differentiation and developmental systems in general.

CONFLICT OF INTEREST

None declared.

ACKNOWLEDGEMENTS

We would like to thank the National Institute of Health for their generous support of our work through the NIH New Innovator Award DP2 116520.

REFERENCES

[1]   Bain G, Kitchens D, Yao M, *et al*. Embryonic stem cells express neuronal properties *in vitro*. DevBiol 1995; 168: 342-357.
[2]   D'Amour K, Agulnick AD, Eliazer S, *et al*. Efficient differentiation of human embryonic stem cells to definitive endoderm. Nat Biotech 2005; 23:1534-1541.

[3]   Johansson B, Wiles M. Evidence for involvement of activin a and bone morphogenetic protein 4 in mammalian mesoderm and hematopoietic development. Mol Cell Biol 1995; 15:141-151.

[4]   Viswanathan S, Benatar T, Rose-John S, *et al.* Ligand/Receptor signaling threshold (LIST) model accounts for gp130-mediated embryonic stem cell self-renewal responses to LIF and HIL-6. Stem Cells 2002; 20:119-138.

[5]   Viswanathan S, Zandstra P. Towards predictive models of stem cell fates. Cytotechnology 2003; 41:75-92.

[6]   Prudhomme W A, Duggar KH, Lauffenburger DA. Cell population dynamics model for deconvolution of murine embryonic stem cell self-renewal and differentiation responses to cytokines and extracellular matrix. Biotech Bioeng 2004a; 88:264-272.

[7]   Prudhomme W, Daley GQ, Zandstra P, *et al.* Multivariate proteomic analysis of murine embryonic stem cell self-renewal *versus* differentiation signaling. Proc Natl Acad Sci U S A 2004; 101:2900-29005.

[8]   Woolf PJ, Prudhomme W, Daheron, L, *et al.* Bayesian analysis of signaling networks governing embryonic stem cell fate decisions. Bioinformatics 2005; 21:741-753.

[9]   Willerth SM, Sakiyama-Elbert SE. Kinetic analysis of neurotrophin-3-mediated differentiation of embryonic stem cells into neurons. Tissue Engineering: Part A 2009; 15:307-318.

[10]   Chickarmane V, Troein C, Nuber UA, *et al.* Transcriptional Dynamics of the Embryonic Stem Cell Switch. PLoSComput Biol 2006; 2:e123-e136.

[11]   Chickarmane V, Peterson C. A computational model for understanding stem cell, trophectoderm and endoderm lineage determination. PLoS ONE 2008; 3:e3478-e3475.

[12]   MacArthur B, Please CP, Oreffo ROC. Stochasticity and the molecular mechanisms of induced pluripotency. PLoS ONE 2008; 3:e3086-e3097.

[13]   Segal, E.Module networks: identifying regulatory modules and their condition-specific regulators from gene expression data. Nat Genet 2003; 34:166-176.

[14]   Shen-Orr SS, Milo R, Mangan S, *et al.* Network motifs in the transcriptional regulation network of Escherichia coli. Nat Genet 2002; 31:64-68.

[15]   Blais A, Dynlacht BD.Constructing transcriptional regulatory networks.Genes Dev2005; 19:1499-1511.

[16]   Vinje WE, Gallant JL.Sparse coding and decorrelation in primary visual cortex during natural vision. Science 2000; 287:1273-1276.

[17]   DeWeese M, Wehr M, Zador AM.Binary spiking in auditory cortex. J Neuroscience 2003 23:7940-7949.

[18]   Olshausen B, Field D. Sparse coding of sensory inputs. Cur OpinNeurob 2004; 14:481-487.

[19]   Shmulevich I, Dougherty ER, Zhang W. Gene Perturbation and Intervention in Probabilistic Boolean Networks. Bioinformatics 2002; 18:1319-1331.

[20]   Hartemink AJ, Gifford DK, Jaakkola TS, *et al.*Combining location and expression data for principled discovery of genetic regulatory network.Pacific Symp Biocomput 2002; 7:437-449.

[21]   Wagner A.How to reconstruct a large genetic network from n gene perturbations in fewer than n^2 easy steps. Bioinformatics 2001; 17:1183-1197.

[22]   Tegner J, Yeung MKS, Hasty J, *et al.*Reverse engineering gene networks: integrating genetic perturbations with dynamical modeling.ProcNatlAcadSci U S A 2003; 100:5944-5949.

[23] Bansal M, Gatta GD, Di Bernardo D, *et al.*Inference of gene regulatory networks and compound mode of action from time course gene expression profiles. Bioinformatics 2006; 22:815-822.

[24] Yeung MKS, Tegner J, Collins JJ.Reverse engineering gene networks using singular value decomposition and robust regression.ProcNatlAcadSci U S A 2002; 99:6163-6168.

[25] Ljung, L. System Identification: Theory for the user. Upper Saddle River, NJ: Prentice Hall PTR; 1999.

[26] Banerjee I, Maiti S, Parashurama N, *et al.* An integer programming formulation to identify the sparse network architecture governing differentiation of embryonic stem cells. Bioinformatics 2010; 26:1332-1339.

[27] Papadimitriou CH, Steiglitz K. Combinatorial Optimization: Algorithms and Complexity. Mineola, NY: Dover;1998.

[28] Yao, X. (Ed). Evolutionary Computation: Theory and Applications. Singapore: World Scientific Publishing; 1999.

[29] Cho CH, Parashurama N, Park EYH, *et al.*Homogenous differentiation of hepaocyte-like cells from embryonic stem cells: applications for the treatment of liver failure. FASEB J 2008; 22:898-909.

[30] Kim SK, Melton DA.Pancreas development is promoted by cyclopamine, a hedgehog signaling inhibitor. Proc Natl Acad Sci U S A 1998; 95:13036-13041.

[31] Habener JF, Kemp DM, Thomas MK.Minireview: Transcription regulation in pancreatic development. Endocrinology 2005; 146:1025-1034.

[32] Kawahira H, Ma NH, Tzanakakis ES, *et al.* Combined activities of hedgehog signaling inhibitors regulate pancreas development. Development 2003; 130:4871-4879.

[33] Chen JK, Taipale J, Cooper MK, *et al.* Inhibition of hedgehog signaling by direct binding of cyclopamine to smoothened. Genes Dev 2002; 16:2743-2748.

[34] White P, May CL, Lamounier RN, *et al.* Defining pancreatic endocrine precursors and their descendants. Diabetes 2008; 57:654-668.

[35] Ben-Shushan E, Marshak S, Shoshkes E, *et al.*A pancreatic β-cell-specific enhancer in the human Pdx-1 gene is regulated by HNF-3β, HNF-1α, and SPs transcription factors. J Biol Chem 2001; 276:17533-17540.

[36] Holland AM, Hale MA, Kagami H, *et al.*Experimental control of pancreatic development and maintenance.Proc Natl Acad Sci U S A 2002; 99:12236-12241.

[37] Soria, B. *In vitro* differentiation of pancreatic β-cells. Differentiation 2001; 68:205-219.

[38] Gradwohl G, Dierich A, LeMeur M., *et al.*Neurogenin3 is required for the development of the four endorine cell lineages of the pancreas.Proc Natl Acad Sci U S A 2000; 97:1607-1611.

[39] Smith SB, Watada H, German MS. Neurogenin3 activates the islet differentiation program while repressing its own expression. Mol Endoc 2004; 18: 142-149.

[40] Watada H, Scheel DW, Leung J, *et al.* Distinct gene expression programs function in progenitor and mature islet cells. J Biol Chem 2003; 278:17130-17140.

CHAPTER 2

Causality Reasoning and Discovery for Systems Biology Investigations

Yi Liu[1,2], Hong Yu[1] and Jing-Dong J. Han[1,2,*]

[1]*Institute of Genetics and Developmental Biology, Chinese Academy of Sciences, China and* [2]*Chinese Academy of Sciences Key Laboratory of Computational Biology, Chinese Academy of Sciences-Max Planck Partner Institute for Computational Biology, Shanghai Institutes for Biological Sciences, Chinese Academy of Sciences, 320 Yue Yang Road, Shanghai, 200031, China*

Abstract: In the past decade, technical innovations in systems biology has made it possible to study the activity of genes, pathways, transcription factors, metabolites and epigenetic states *in vitro* or *in vivo* on the genome-wide scale. In the design and analysis of such experiments, researchers often face an imperative question: To what extent and by which means can we extract valuable biological knowledge (which is often embodied as undirected and directed interactions between biological factors) from a particular experiment? In this chapter, we review state of the art algorithms for the structure learning of Bayesian networks and the elucidation of causal knowledge to partially address this question. Specifically, the distinct feature of each algorithm and its connections with other algorithms are highlighted in the context of causality reasoning and discovery for systems biology investigations.

Keywords: Embryonic stem cells, regulatory networks, Bayesian networks, causality, structure learning, probablistic graphical models, interference, exact learning, approximate algorithms, large-scale datasets, genomics and epigenomics.

1. INTRODUCTION

The fast development of high-throughput biochemical assays significantly promotes the progression of systems biology research. As suggested by its name, systems biology studies the molecular interactions of a biological system at a global, systematic level. This is quite different than traditional molecular biology experiments, in which the research focus is usually a single gene and its partner. In system biology investigation, we can obtain global views of a cell at the genome-wide scale from different perspectives using a number of high-throughput experimental techniques. For example, Microarray and RNA deep sequencing

---

*Address corresponding to Jing-Dong J. Han: Chinese Academy of Sciences Key Laboratory of Computational Biology, CAS-MPG Partner Institute for Computational Biology, Shanghai Institutes for Biological Sciences, Chinese Academy of Sciences, 320 Yue Yang Road, Shanghai, 200031, China; E-mail: jdhan@picb.ac.cn

techniques allow us to measure the expression of all genes in a eukaryotic genome; Chromatin- immunoprecipitation followed by microarray (ChIP-chip) or deep sequencing (ChIP-seq) enables the identification of all targets of a transcription factor or all carrier of an epigenetic modification; Yeast two Hybrid (Y2H) technique for the comprehensive profiling of protein-protein interactions and the mass spectroscopy techniques for the analysis of metabolites in a cellular system, just to mention a few.

These high-throughput experiments often yield huge volumes of data, which are typically observational measurements or results of interventional studies. Here, observation means that the experiments are performed in a passive manner, *i.e.,* no intentional alternation of a cellular system took place before an experiment is carried out. For example, a microarray study is carried out without specific treatment of the testing tissue. By contrast, in an interventional study, researchers deliberately exert a certain influence to a cellular system to examine the resultant of this perturbation [1]. Naively thinking, it is possible to identify non-causal interactions between elements in a system by testing their Pearson cross-correlations. And also, it seems to be rather straightforward to disambiguate cause and effects through an interventional study. However, things are more complicated beyond this line of thinking. First, causal and non-causal interactions could be direct or indirect. Is there a principled approach to separate direct interactions with indirect interactions? Without this distinction, more interactions will be found than it should be and those false positive ones will prohibit the correct identification of biological knowledge. Second, not all non-linear interactions can be correctly identified using simple, linear indicators, such as cross correlations. Third, rather than the case of interventional studies, from pure observational experiments, there are still causal relationships that could be identified. However, there is not a simple approach which could identify them correctly. Finally, it is not easy to integrate the information in different types of experiments in a coherent manner. To this end, in this chapter, we introduce a principled mechanism, namely the Bayesian networks (BN), for performing causal reasoning and inference both qualitatively and quantitatively [2].

This chapter is organized as follows. In section 2, we introduce the Bayesian networks (BN) as a causality modeling and inference representation. Specifically, we will focus on presenting the recent advances in learning Bayesian network structures from data, which is the key technique for inferring causal relations from systems biology data. In section 3, we introduce interventional study and the dynamic Bayesian networks representation for resolving the causality ambiguities in learning BN structure. We also introduce how to elicit consistent causal interpretations from prior knowledge and BN structure. The application of BNs to systems biology investigations will be briefly sketched in Section 4. Finally, we conclude this chapter in Section 5.

2. THE STRUCTURAL LEARNING OF BAYESIAN NETWORKS

The Bayesian network formalism has been extensively used for representing and inferring causal relationships in systems biology research [3]. Briefly, a Bayesian network is a direct graphical model in which each node represents an entity which we would like to investigate its causal interaction with other entities (nodes). Here, causality means that for any two interacting nodes, one must be the cause of the other. For this reason, each edge in a BN is associated with an arrow to indicate the direction of information flow. Furthermore, there is no directed cycle in a Bayesian network structure since otherwise the definition of causality would be violated. As a result, the structure of a BN must be a *directed acyclic graph* (DAG), in which we can associate a *topological ordering* of the nodes where all edges are consistent with it to signify the rank of causality [2].

Specifically, all the nodes that directly point towards a node in a BN are defined as the *parents set* of the node. Similarly, we can define the *children set* as the collection of nodes that are pointed from their parent node. The *Markov blanket* of a node is defined as the minimal conditioning set of nodes which separate a node with the rest of nodes in a BN. Under the *faithfulness assumption* (which will be introduced shortly), it can be shown that the Markov Blanket (MB) of a node is the union of its parents, children and all the other parents of its children. The most important notion in Bayesian networks is *family*, which is defined as the union of a node and its parents set. Indeed, the joint probability distribution parameterized by a BN can be factorized as a product of the local conditional probability distributions (local CPD) for the nodes of each family. For each node in discrete BNs, the local CPD is simply a multinomial distribution for each combinatorial configuration of its parents; and for linear Gaussian BNs, the local CPD is defined as a Gaussian distribution whose mean is a linear combination of the values of its parents. Given this factorization, BNs simplify the joint probability distribution greatly due to the localized nature of each CPD [2].

For the structural learning of BNs, the notion of family again plays an important role. First, for the scoring-based structural learning approach, which aims to find the BN structure that best fit to the data while having minimal model complexity, it can be shown that the decomposability of most scoring functions according to the local "family" structures has greatly simplified the graph search procedure. Second, since the factorization of the joint probability distribution based on each family implies a large number of conditional independency relations between sets of nodes, it is possible to learn the BN structure by testing these conditional independency relations. In fact, the scoring based and the statistical tests based approaches are the two major classes of algorithms for the structure learning of BNs. They can also be integrated to design more powerful, hybrid BN learning algorithms, which will be introduced in the next section [4].

2.1. Learning Bayesian Network Based on Statistical Tests

As we have briefly mentioned above, it is possible to learn the Bayesian network structure by testing conditional independency relations based on the training data available. In fact, the well functioning of this approach is based on the *faithfulness assumption*, which states that *if and only if* the conditional independency relations implied by the BN structure are exactly those hold by the data. For a fixed BN structure, the *d-separation* criterion can be used to judge whether a conditional independency relation is encoded by the directed acyclic graph. It works by checking whether the conditioning set can block all paths between two sets of nodes. In terms of "block", cares should be taken in handling the case where the arrows of two flanking edges in the path point towards a node. In fact, if the node itself or any of its descendants is at the conditioning set, this path is NOT blocked at this node. On the other hand, if neither the node nor its descendants are in the conditioning set, the path is blocked otherwise. This is contradictory to conventional definition of "block". However, the aforementioned *"explain away"* semantics is very important for correct reasoning with causality [5].

With this background knowledge, we can understand how statistical tests based approaches function in learning BN structures. In technical essence, each positive result of an independency test will impose a constraint to the BN structure. Therefore, such algorithms are often called "constraints-based" BN structural learning algorithms. In chronological order, the most important algorithms in this class are the IC [6], PC algorithm and variants of the PC algorithm [7]. All these approaches work in a similar manner. First, they learn the *undirected skeleton* of the BN, which is essentially the backbone of BN, formed by discarding the arrows of the edges in the true directed acyclic graph. Then, in the second step, they establish the directionality of these edges to generate a complete BN structure. In the first step, statistical tests are used to judge whether an edge exist between a pair of nodes by checking if there exists a conditioning set which is able to separate their dependency. If there is no way to separate the node pair, then there must be an edge between them. In the IC algorithm [6], these tests are conducted in an exhaustive manner. However, this is not necessary since many of them convey redundant information. In the PC algorithm [7], we start with a fully connected adjacency graph. Then, statistical tests are performed with increasing cardinalities of conditioning sets from the adjacency (parents/children) set of each node. It can be shown that if two nodes are always tested to be dependent if the conditioning set traverses all subset of the adjacency sets of the two nodes respectively, there must be an edge between them. As a result, the correctness of the PC algorithm is guaranteed, although a large number of conditioning sets are not used in the statistical tests.

In the second step of the IC and PC algorithm, the main task is to establish the directions of edges in a BN. However, cares should be taken in this step since there is no way to distinguish BNs in an equivalent class based on the encoded conditional independency relations. That is, only the direction of *compelled edges* can be determined since they are invariant within the equivalence class, the direction of non-compelled edges are unresolved [8, 9]. Furthermore, it can be shown that the set of *v-structures* in the DAG is representative of an equivalence class of BNs, where a *v-structure* is defined as a triplet of nodes $a->b<-c$ with no edge exists between a and c. As a result, our task here is reduced to identifying all *v-structures* in the BN skeleton. To this end, we make use of the results of the conditional independency tests in the first step again: The *v-structures* are essentially the set of unshielded triplets *a-b-c* (unshielded means there is no edge between *a* and *c*) where b is not in the conditioning set of *{a, c}*. Based on the identified v-structures, we can applying Meek's rule to orient the other compelled edges to obtain a class of equivalent DAGs, or to orient all the undirected edges to obtain a DAG in the equivalent class [10].

It is possible to further reduce the number of conditional independency tests in learning the BN structure proposed a new algorithm which recursively reduce the problem of learning the BN structure for a large domain to two smaller BN learning problems [11]. In this divide and conquer approach, many unnecessary conditional statistical tests can be avoided since *moral edges* (which connect different parents of a node) are eliminated early in the small graphs. Specifically, vertex set decomposition in this algorithm is performed by first learning an undirected independence graph and decompose the graph into several parts by running the junction tree algorithm [12]. At the leaves of this decomposition, the BN skeletons of the small vertex sets are constructed using either the PC or IC algorithm. In this process, the separator sets (conditioning sets) generated by the junction tree and the PC/IC algorithm are saved. In the subsequent graph combining stage, all the edges of two smaller graphs are included in the combined graph except those inconsistent ones in their common vertices. Finally, when the global BN skeleton of the domain variables is obtained, v-structures can be identified using the same approach as the PC/IC algorithm. Again, Meek's rule [10] can be employed to construct a Partially Directed Acyclic Graph (PDAG, which represents a class of equivalent DAGs) or a DAG.

Now, we have briefly sketched the constraints-based BN learning algorithms which are based on statistical tests. The performance of these algorithms is mainly confined by the accuracy of the statistical tests, which could be a serious problem when the number of train cases is limited. To improve the quality of the learning results, the scoring-based scheme is proposed to learn Bayesian network structures, which will be introduced in the next section.

2.2. The Scoring-based Approach to Bayesian Network Learning

The basic idea of the scoring-based approach to learning Bayesian networks is to find the BN structure, which has the best fitness to data while having the least number of parameters at the same time [4]. To make a compromise between the two terms, the following approaches are frequently used to define the scoring function: the Akaike Information Criterion (AIC) [13], the Bayesian Information Criterion (BIC) (14) and the Minimum description length principle (MDL) [15]. Specifically, the AIC is derived from the frequentist's viewpoint in measuring the quality of the bias-variance tradeoff of a statistical model, while the BIC is derived from a Bayesian viewpoint for selecting the best statistical model over a number of candidates. In particular, the AIC criterion penalizes the fitness of the model with exactly the number of its parameters, while the penalization term in the BIC criterion is $0.5 \ln N$ fold larger. Since BN structure learning task is essentially a model selection problem, the BIC criterion might be more appropriate in this context.

The underlying assumption behind the *minimum description length* (MDL) principle is somewhat different [15]. Briefly, it seeks the most space-efficient way to encode the data using a statistical model, in which two space requirements are needed for data compression, one is to define the model and the second is to encode the data using the model. If the model is too simple, there will not be much space saving after the data is encoded. On the other hand, if the model is too complex, the data can be compressed much deeper but more parameters are needed to define the model. Therefore, the MDL principle also defines the fitness *vs.* model complexity tradeoff, which is equal to the BIC criterion under some assumptions in learning BN structures.

The problem with AIC, BIC and MDL scoring metrics is that they are only approximations when we the number of training data is finite. To be more precise, one can derive the exact *Bayesian* scoring function [16, 17] for a BN structure, as we will introduce below.

Here, we only describe the derivation of the *Bayesian* scoring metric for discrete BNs [18]. In this case, we assume that each variable in a BN follows a multinomial distribution given the full assignment of its parents, and the parameters of such distributions are independent for different nodes and parents' assignments. Furthermore, we also assume the prior distribution of the multinomial parameters in a conditional density function only depend on the local graph topology, *i.e.,* which nodes are the parents of a particular node, and there priors are parameterized by *Dirichlet* distributions. Based on these assumptions together with the completeness of data, the conjugacy of *Dirichlet* with multinomial distributions, the Bayesian metric can be derived with a closed-form solution [18].

There is, however, an important issue that has to be addressed properly, namely the hypothesis equivalence [8]. As we have introduced earlier, a BN encodes a number of conditional independence relations which can be read out from the graph using the d-separation rule, and two BNs are said to be *equivalent* if they represent the same set of conditional independencies. The hypothesis equivalence implies that in the causality discovery process, there is no way to distinguish two equivalent BNs, that is, their graph prior and the likelihood of data given the graph should be identical. Here, likelihood equivalence dictates that the prior Dirichlet distribution over the parameters of the joint multinomial distribution over domain variables is identical for two equivalent DAGs. To fulfill this requirement, cares should be taken in specifying the prior distributions of the multinomial parameters in local CPDs.

It can be shown that a joint Dirichlet distribution over all domain variables can be used to define the prior parameter distribution for any BN, and this definition scheme satisfies likelihood equivalence [16]. In fact, such a joint prior distribution can be parameterized by a fully connected BN, where the local CPDs are also Dirichlet distributions and the parameter independence properties are naturally satisfied. Based on this finding, the prior distributions of the CPDs in an arbitrary BN can be defined in the following way: First, for each CPD, we reorder the nodes of a fully connected BN so that the family of this CPD appears ahead. Then, the local CPD distributions of this reordered BN can be computed by a change of variables from the original fully connected BN. This is because all fully connected BNs are equivalent due to likelihood equivalence. Finally, the parameters of the CPD in the reordered BN can be used to define the corresponding CPD in the original BN, since the prior Dirichlet distribution for a local CPD only depends on the parents of the node [16].

With this likelihood-equivalence definition of prior probability, the posterior probability of the data given a BN (which is termed as the Bayesian Dirichlet equivalent metric, *i.e.,* BDe metric) can be derived by integrating out the multinomial parameters [16]. It can be shown that the BDe metric is a special case of the Bayesian metric where the prior distribution for each local CPD is determined by a joint Dirichlet distribution. As we have shown above, this constraint is essentially posed by the likelihood equivalence property of BNs: Only with this constraint, it is provable that the BDe metric is invariant under covered-edge reversals, where the likelihood equivalence criterion is satisfied naturally [9].

The above discussion is mostly focused on discrete BNs. For linear-Gaussian BNs, following the same line of reasoning, a counterpart of the BDe metric, the Bayesian Gaussian equivalence (BGe) metric can be derived. Here, the Multivariate Gaussian *vs.* Normal-Wishart conjugacy has replaced the Dirichlet *vs.* multinomial conjugacy. Nevertheless, the likelihood equivalence again plays a central role in mathematical derivation [17].

Now we have introduced several ways about designing a scoring metric to quantify the goodness of fit of a BN structure with training data. The only thing left is to find a way to search for the best scoring BN structures among all possible directed acyclic graphs. However, it turns out that the number of DAGs is super-exponential to the number of nodes in the domain, which implies that a naïve exhaustive enumeration approach does not simply work [19]. To overcome this problem, the simplest approach is to use a greedy ascent search, where only a local graph change is performed as each step [4].

Specifically, the greedy ascent search algorithm starts with an empty graph or a prior BN structure. In each step, it considers all valid edge insertions, deletions and reversals which will end up with a valid DAG. Then, the operation with the largest score increase is performed (ties are breaking randomly), ended with a DAG which is only locally different. This is done repeatedly until the score of the BN reaches a local maximum [16].

In practice, the greedy ascend search algorithm works fairly well: the local optimal solution is often close to the exact best DAG. To further improve the quality of heuristic search, two ways are often exploited. First, the search could run multiple times with a random BN as the start point and the DAG with the largest score is output as the result. Second, TABU strategy can be used in the greedy ascend search [20]. Here, "TABU" means forbidden as this strategy prohibits the current search from undoing a recently performed operation. It can be shown that by allowing the score decrease temporarily during the heuristic search process, the "TABU" strategy is able to bypass small local optima and obtain better solutions.

Note that at each local search step, we have to first enumerate all valid graph operations which change the current DAG to a new DAG, and evaluate the change of the score for each of these operations. Note that most scoring metrics, including AIC, BIC, MDL and BDe, can be decomposed into a sum of scores for each family in a DAG. Therefore, we only need to compute the score changes of the affected families for each graph operation. This computational process can be speed up further by caching the score changes for all graph operations of the current DAG, since for the new DAG derived from the original DAG with a single graph operation, most of its score changes need not re-computed: we can just use the cached values as the changes of local graph topologies are identical. We only need to update the score changes for the families with altered edge connections between the old and new DAGs. Using these two strategies, the speed of graph search can be lifted by two orders of magnitude in terms of the number of domain variables [21].

The main problem with greedy ascent search procedures is that there is no warrantee about the accuracy of the solution. Fortunately, it has recently been shown that by using well designed dynamical learning procedures, exact BN structural learning is even tractable with medium sized problems [19, 22, 23].

First, similar to greedy ascent search, our task is to find the BN structure with optimal score, which is essentially a *maximum a posteriori* (MAP) problem. For this problem, the exact learning algorithm works by recursively computing the best sink (the lowest node in the topology ordering) of the network [23]. Specifically, we first compute the scores of each node for all combinations of its parents' sets. Note that the exponential exposition here can be avoided by setting the maximal cardinality of a family. Based on this result, we can then select the best parents set for each node for all topological orderings of the domain. Finally, with increasing cardinality, the best BN structure can be constructed for all possible subset of nodes in the domain by testing which {sink, smaller network} combination maximizes the score of the current subset. Of course, the best BN structure for the largest subset (*i.e.,* the domain variables) is our desired output [23].

Second, if we are interested in computing the exact Bayesian posterior probability for each specific graph motif, *e.g.,* a directed edge, the contribution from all BN structures should be considered. The results obtained in this *Bayesian model averaging* formulation are more robust than the previous *MAP* learning scheme. Of course, this is a more challenging task. However, it can also be solved efficiently using dynamic programming techniques [19, 22].

3. LEARNING AND INTERPRETING WITH CAUSALITY

In the previous section, we have introduced the Bayesian network formalism for mining causal knowledge from observational data (although how to make proper causal interpretations will be postponed to section 3.4). However, in some cases, we can detect stronger causal relationships from the systems biology experiments which generated through an *interventional* study.

3.1. Interventional Studies

For the term *intervention*, we mean an external perturbation is exerted to a system, in a way that by observing the change of the states of the system, we are able to identify which factors are causally responsive to this perturbation [1]. Note that cares should be taken in learning BN structure from interventional data, since different types of intervention often have very different impacts to the system, which requires the learning algorithm to distinguish and handle them properly.

In a recent work [24], the scoring-based BN learning approach based on the BDe metric is modified to handle four different types of intervention: 1) No intervention; 2) Perfect Intervention [1]; 3) Imperfect intervention, including "switching parents" intervention [25], "unreliable" intervention [26] and "soft" intervention [27]; 4) Uncertain intervention [24]. Here, No intervention means that the training data are pure observational; perfect intervention means that the intervention deterministically set the states for a node in the BN, which effectively "cut off" the edges from the parents to this node. For imperfect intervention, we only know that the distribution of the states of the target node changes, but its value is not clamped deterministically. This includes the special case that the functional dependency between the node and its parents changes under intervention ("switching parents" intervention); the special case where the intervention has a probability to fail ("unreliable" intervention) and the special case where the intervention just increases the probability that the node is in a certain state ("soft" intervention), which is reflected as a change of the hyper-parameters in the BDe metric. Finally, the target nodes are not known for uncertain interventions. To tackle with this problem, we can augment each uncertain intervention as an extra node and then perform constraint BN structure learning in the enlarged domain. Here, the edges in a BN are only allowed to point from interventional nodes to actual nodes or between actual nodes.

3.2. Hybrid BN Structure Learning Algorithms

In section 2, we have introduced two major classes of BN structure learning algorithms, namely, the constraints-based and the scoring-based approaches. The former is completely based on the results of statistical tests. As a result, many false positive edges may appear in a BN due to limited power of such tests, especially when the conditioning set is large. Besides, for the same reason, the inferred edge directions are not very stable. On the other hand, if the number of variables in a domain is large, the speed of scoring-based approaches will decrease quickly and the structural search is likely to entering incorrect regions. In this section, we show these two classes of algorithms can be integrated into accurate and scalable hybrid learning algorithms.

The basic idea of hybrid learning algorithm is to use the intermediate results of constraint based learning algorithms, that is, the undirected skeleton of the BN structure, to constrain the search space of scoring-based learning methods [28]. Note that the spurious edges in the skeleton no longer become a problem in hybrid learning algorithms, since scoring-based algorithms are able to find close to optimal solutions within the skeleton, which has hopefully eliminated most of the false positive edges. On the other side, the undirected skeleton has vastly reduced the search space, which speeds up the scoring based BN learning algorithms dramatically.

One notable example of hybrid BN structure learning approach is the MMHC algorithm, which essentially uses a different strategy to learn the skeleton of a BN [28]. Briefly, It maintains a candidate parents/children (CPC) set for each node, and gradually add nodes to this set in the following way: For each variable *not* in this set, the algorithm computes its *minimum* association with the node by conditioning on all possible subsets of the CPC set. Then, the variable with the *maximum "minimum association"* is selected and added to the CPC set. This max-min strategy is very effective for finding relevant nodes rapidly, but at the risk of adding more false positive nodes in the CPC set at the beginning. Of course, combining the IC/PC/vertex set decomposition algorithms with scoring-based learning algorithms are also valid approaches to designing hybrid BN structure learning algorithms.

3.3. Learning Dynamic Bayesian Networks

In systems biology investigations, many experiments are conducted along a series of well planned time points. For example, to understand how gene expressions are well coordinated over a developmental process, it is possible to perform RNA-seq experiments at several time points for each developmental stage. Based on the fact that the cause of a result must occur ahead of it, it is possible to constrain the directions of causality using time information. Therefore, we may elicit more knowledge about causality by taking account of such constraints. For this purpose, a variant of Bayesian network representation, the *Dynamic Bayesian Network (DBN)*, is developed to characterize the dependencies of a set of factors over time series [29].

To simplify the structure learning problem, we often assume the interdependencies of the factors in any two consecutive time frames are identical, which implies that the DBN is *time homogenous*. Based on this simplification, the structure of a DBN can be represented using two parts: the prior network as well as the transition network. The former component is essentially an ordinary BN, which characterizes the stationary dependencies of the factors in the domain before the initialization of the time series. Therefore, it is usually not very interesting and we can infer its structure by simply applying a conventional BN structure learning algorithm to the data items before the start point. On the other side, the transition network represents the important interdependencies which are invariant over any two consecutive time frames. As such, the main task of learning DBN structures is to infer the transition network [29].

Specifically, the edges in a transition network can be divided into two parts: the intra- and inter-slice edges. For the transition network at time t and $t+1$, an intra-slice edge connects two nodes at time $t+1$, while an inter-slice edge starts from a node at t but ends at a node at time $t+1$. By unfolding and tiling up the transition networks repeatedly, the long chain like structure of a DBN can be restored exactly [29].

It can be shown that the structure of the transition networks can also be learnt in a relative simple manner. First, the instantiations of domain variables at two consecutive time points are concatenated to form long vectors with doubled sizes. In other words, each variable is spitted to two variables, indicating the "previous time" and the "current time" instantiations. Second, constrained BN structure learning are performed on this enlarged domain. Specifically, we only allow edges link from nodes at "previous time" towards nodes at the "current time" (inter-slice edges) or edges connect two nodes at the "current time" (intra-slice edges). Of course, during the structure search process, it is necessary to keep the acyclicity of the transition network, as in ordinary BN structure learning algorithms [29].

3.4. Making Proper Causal Interpretations

As we have mentioned before, Bayesian networks can be divided into equivalence classes based on the set of conditional independencies they encode. Without any prior knowledge about BN structure, it is not possible to distinguish two BNs in an equivalence class based on the set of conditional independencies embodied in the training data. As a result, only compelled edges in a BN (whose directions are fixed in an equivalence class) can be interpreted as causal interactions, while the directions of non-compelled edges are unresolved. To distinguish these two classes of edges in a BN, we can use an efficient algorithm developed by Chickering based on covered edge reversals [9].

Things could be more complicated when the directions of some edges are known *a priori* or constrained. For example, to make causal explanation for DBNs, the directions of inter-slice edge connections are fixed, which eliminate many equivalent BN structures inconsistent with them. As a result, many non-compelled edges could become compelled due to the introduction of these constraints. To account for this, Meek propose four rules R1 to R4 to characterize the propagation of such constraints along the BN structure, which can be used to infer the set of compelled edges in a BN under constraints [10]. However, when the BN structure is free, Chickering's algorithm is more preferred since it is computationally much faster [9].

4. APPLICATIONS OF BAYESIAN NETWORKS TO SYSTEMS BIOLOGY

In the earlier stage of system biology, the application of Bayesian networks in microarray data is rather limited, since the number of measurements is often far from enough given the large number of genes in a genome. However, with the rapid development of new high throughput technologies, such as the next generation sequencing facilities, it is possible to deep sequence personalized human genomes, which greatly mitigated the bottleneck of data scarce. On the other hand, the emergence of ChIP-seq technologies has enabled genome wide profiling of protein-DNA, protein-RNA interactions, histone modifications, inter- and intra-chromosomal interactions and other kinds of phenomenon comprehensively. Furthermore, Genome-Wide Association Studies (GWAS) also have provided the phenotype and genotype data for thousands of individuals. The fast emergence of large-scale biological data has made it possible to apply Bayesian network to discover the underlying biological knowledge. For example, in ChIP-seq experiments, there are just tens or hundreds of factors being investigated (DNA-binding proteins, histone modifications, *etc.*), but the number of observation is abundant (tens of thousands of genes). Therefore, it is possible to identify the regulatory relationship between them. For example, in embryonic stem (ES) cells, histone H3 lysine 27 trimethylation (H3K27me3) are highly enriched, its deposition and recognition are carried out by the PRC2 complex and the PRC1 complex (both are polycomb group (PcG) proteins), respectively [30]. By applying BN on ChIP-chip profiles of H3K27me3, two PRC2 proteins (EED and Suz12) and a PRC1 protein (Rnf2) in mouse ES cells [31], we have recaptured the sequential binding and recruitment relationships between different PcG proteins and the histone H3 lysine 27 trimethylation (H3K27me3) catalyzed or recognized by the PcG proteins from ChIP-chip data (Fig. **1**) [32]. As another example, in cancer stem cells, genes that are marked by H3K27me3 in normal stem cells are frequently marked by H3K9 di- and tri-methylations [33]. The BN inferred from human histone modification ChIP-seq data identifies H4K20me3, a modification involved in DNA repair, may synergistically convert H3K27me3 to H3K9me3 and H3K9me2, therefore implicating a role of DNA repair in the transition of normal stem cell epigenetic signatures to cancer stem cell signatures [32].

For GWAS, which measures the SNP and gene expression of thousands of people, Bayesian network and many other probabilistic modeling approaches can be employed to pinpoint functional SNPs and discovering relevant regulatory relationships [34]. Even for the microarray data which was not sufficient for Bayesian network analysis, the reduced cost has made it possible to perform thousands of measurements over different individuals, which greatly increased the practicability of applying Bayesian network to this problem. Although this situation has greatly improved, it is still nearly impossible to identify the regulation of tens of thousands of genes *de novo* from just thousands of observations. However, prior knowledge and other information could be used to filter implausible causal interactions before the application of Bayesian networks. We believe with the fast development and the cost reduction of the currently available as well as new high throughput experimental techniques, the application scope of Bayesian networks will be greatly expanded and the accuracy of discoveries will also improve in the near future.

5. CONCLUDING REMARKS

In this chapter, we have reviewed the complete causality reasoning and discovery process in the Bayesian networks formalism. In particular, we cover the statistical tests based and scoring based BN structure learning algorithms in great detail. We also discuss how to resolve causality ambiguity by incorporating prior knowledge, performing interventional studies and using time constraints. The issue of making proper causal interpretation based on the learning results is also sketched briefly. Finally, we illustrate many plausible applications of BNs in systems biology. It is hoped that this in-depth review will help the readers in biology aware of this powerful data mining tool and help the readers with a strong computational background familiar with the basic ideas of many important BN learning algorithms in a few hours.

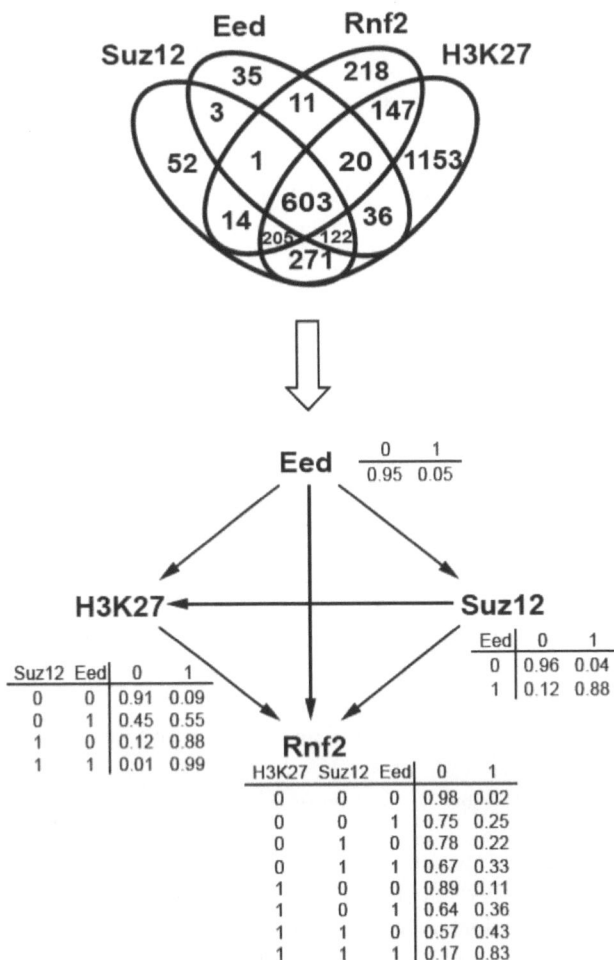

Figure 1: This figure illustrates the application of Bayesian network on ChIP-chip data. The Venn diagram shows the number of overlapped target genes of DNA-binding proteins and a histone modification. From the Venn diagram, the only thing learned is that the four factors share a significant number of target genes. However, it was known that the four factors have causal relationships among them, that is, the polycomb group (PcG) proteins Ezh2 and Suz12 first interact with each other, then they can catalyze the formation of histone H3 lysine 27 trimethylation (H3K27me3) at their binding sites, which in turn can recruit another PcG protein Rnf2. The Bayesian network inferred from ChIP-chip data which mapped the PcG binding and H3K27me3 at all mouse promoters can recapture all the known relationships found by genetic and biochemical studies. The tables show the conditional probability distributions of the status of a child node given the status of its parent node (s). The figure is adapted from [32].

CONFLICT OF ITNEREST

None declared.

ACKNOWLEDGEMENTS

This work was supported by grants from the China Natural National Science Foundation (Grant #30890033 and 91019019), Chinese Ministry of Science and Technology (Grant #2011CB504206) and Chinese Academy of Sciences (Grant #KSCX2-EW-R-02 and KSCX2-EW-J-15) and stem cell leading project XDA01010303 J.D.J.H.

REFERENCES

[1]   Pearl J. Causality: models, reasoning, and inference. Cambridge University Press 2000.
[2]   Koller D, Friedman N. Probabilistic Graphical Models: Principles and Techniques. MIT Press 2009.
[3]   Friedman N, Linial M, Nachman I, Pe'er D. Using Bayesian networks to analyze expression data. J Comput Biol 2000; (7):601-20.
[4]   Heckerman D. A Tutorial on Learning with Bayesian Networks. Learning in Graphical Models, M Jordan, ed MIT Press 1999.
[5]   Pearl J. Probabilistic Reasoning in Intelligent Systems: Networks of Plausible Inference. Morgan Kaufmann Publishers 1988.
[6]   Pearl J, Verma T. A Theory of Inferred Causation. Proceedings of the 2nd International Conference on Principles of Knowledge Representation and Reasoning (KR'91) 1991:441-52.
[7]   Spirtes P, Glymour C, Scheines R. Causation, Prediction, and Search, 2nd ed. MIT Press 2000.
[8]   Verma T, Pearl J. Equivalence and synthesis of causal models. Proceedings of the Sixth Annual Conference on Uncertainty in Artificial Intelligence 1991:255-70.
[9]   Chickering DM. A Transformational Characterization of Equivalent Bayesian Network Structures. Proceedings of the Eleventh Annual Conference on Uncertainty in Artificial Intelligence 1995:87-98.
[10]   Meek C. Causal inference and causal explanation with background knowledge. Proceedings of the Eleventh Annual Conference on Uncertainty in Artificial Intelligence 1995:403-10.
[11]   Xie X, Geng Z. A recursive method for structural learning of directed acyclic graphs. J Machine Learning Res 2008; 9:459-83.
[12]   Lauritzen SL, Spiegelhalter DJ. Local Computations with Probabilities on Graphical Structures and their Application to Expert Systems. J Royal Statist Soc Ser B (Methodological) 1988; 50 (2):157-224.
[13]   Akaike H. A new look at the statistical model identification. IEEE Transactions on Automatic Control 1974; 19 (6):716-23.
[14]   Schwarz GE. Estimating the dimension of a model. Annals of Statistics 1978; 6 (2):461-4.

[15]   Grünwald P. The Minimum Description Length principle. MIT Press 2007.
[16]   D.Heckerman, Geiger D, Chickering DM. Learning Bayesian Networks: The Combination of Knowledge and Statistical Data. Machine Learning 1995; 20 (3):197-243.
[17]   D.Geiger, Heckerman D. Learning Gaussian Networks. Proceedings of the Tenth Annual Conference on Uncertainty in Artificial Intelligence 1995:235-43.
[18]   Heckerman D, Geiger D, Chickering DM. Learning Bayesian Networks: The Combination of Knowledge and Statistical Data. Machine Learning 1995; 20 (3):197-243.
[19]   Koivisto M, Sood K. Exact Bayesian Structure Discovery in Bayesian Networks. Journal of Machine Learning Research 2004; 5:549-73.
[20]   Cvijovic D, Klinowski J. Taboo search - an approach to the multiple minima problem. Science 1995; 267:664-6.
[21]   Friedman N. Learning Belief Networks in the Presence of Missing Values and Hidden Variables. Proceedings of the Fourteenth International Conference on Machine Learning 1997:125-33.
[22]   Koivisto M. Advances in Exact Bayesian Structure Discovery in Bayesian Networks. Proceedings of the 22nd Conference in Uncertainty in Artificial Intelligence 2006.
[23]   Silander T, Myllymäki P. A Simple Approach for Finding the Globally Optimal Bayesian Network Structure. Proceedings of the 22nd Conference in Uncertainty in Artificial Intelligence 2006.
[24]   Eaton D, Murphy K. Exact Bayesian structure learning from uncertain interventions. Proc 12th International Workshop on Artificial Intelligence and Statistics 2007:107-14.
[25]   Tian J, Pearl J. Causal Discovery from Changes. Proceedings of the 17th Conference in Uncertainty in Artificial Intelligence 2001:512-21.
[26]   Korb KB, Hope LR, Nicholson AE, Axnick K. Varieties of Causal Intervention. 8th Pacific Rim International Conference on Artificial Intelligence 2004:322-31.
[27]   Markowetz F, Grossmann S, Spang R. Probabilistic soft interventions in Conditional Gaussian networks. Proc 10th International Workshop on Artificial Intelligence and Statistics 2005.
[28]   Tsamardinos I, Brown LE, Aliferis CF. The max-min hill-climbing Bayesian network structure learning algorithm. Machine Learning 2006; 65 (1):31-78.
[29]   Friedman N, Murphy KP, Russell SJ. Learning the Structure of Dynamic Probabilistic Networks. Proceedings of the Fourteenth Conference on Uncertainty in Artificial Intelligence 1998:139-47.
[30]   Sparmann A, van Lohuizen M. Polycomb silencers control cell fate, development and cancer. Nat Rev Cancer 2006; 6 (11):846-56.
[31]   Boyer LA, Plath K, Zeitlinger J, *et al.* Polycomb complexes repress developmental regulators in murine embryonic stem cells. Nature 2006; 441 (7091):349-53.
[32]   Yu H, Zhu S, Zhou B, Xue H, Han JD. Inferring causal relationships among different histone modifications and gene expression. Genome Res 2008; 18 (8):1314-24.
[33]   Ohm JE, McGarvey KM, Yu X, *et al.* A stem cell-like chromatin pattern may predispose tumor suppressor genes to DNA hypermethylation and heritable silencing. Nat Genet 2007; 39 (2):237-42.
[34]   Tang W, Wu X, Jiang R, Li Y. Epistatic module detection for case-control studies: a Bayesian model with a Gibbs sampling strategy. PLoS Genet 2009; 5 (5):e1000464.

CHAPTER 3

Exploring Stem Cell Gene Expression Signatures using AutoSOME Cluster Analysis

Aaron M. Newman and James B. Cooper[*]

University of California, Santa Barbara, USA

Abstract: Stem cell laboratories around the world routinely generate whole-genome expression data to study systems-level processes in stem cell biology, and computational clustering methods are critical for the genome-wide analysis of such large data sets. To address major limitations with commonly used clustering approaches, we developed a novel computational method called AutoSOME to automatically cluster large, high-dimensional data sets, such as whole-genome microarray expression data, without prior assumptions about cluster number or data structure. In previous work we demonstrated that AutoSOME clustering is an effective method for studying genome-wide expression patterns in stem cells. Here we present a primer that describes how to use this method to perform comprehensive cluster analyses of stem cell gene expression data. We include two detailed protocols illustrating the identification of gene co-expression modules and clusters of cellular phenotypes in a single step (Protocol 1), and the visualization of transcriptome variation among stem cells using an intuitive network display (Protocol 2). The workflow described in this chapter is sufficiently general for use with a wide variety of in-house and publicly available genomics data sets.

Keywords: Gene clustering, whole-genome expression data, cellular phenotypes, AutoSOME gene co-expression modules, machine learning, cartography, graph theory.

INTRODUCTION

Stem cells have significant potential for elucidating fundamental mechanisms of developmental and disease biology, and are widely believed to hold great promise for regenerative medicine. In recent years, activity in stem cell research has greatly accelerated, owing largely to the advent of cellular reprogramming [1], an increase in funding (see [2] and [3]), and the use of systems-level technologies to

---

*Address correspondence to James B. Cooper: Biomolecular Science and Engineering Program, and Molecular, Cellular and Developmental Biology Department, University of California, Santa Barbara, CA 93106, USA; Tel: (805) 893-8028; Fax: (805) 893-4724; E-mail: jcooper@lifesci.ucsb.edu

characterize the pluripotent state (*e.g.,* [4-9]). For example, since the successful generation of induced pluripotent stem cells (iPSCs) from mouse fibroblasts in 2006 [1], increasingly effective strategies have been devised for creating iPSCs from various progenitor cells (*e.g.,* [10-12]), and viable mice were born from iPSC-derived embryos [13]. Important insights have also been made with regard to similarities and differences in gene expression [6, 14-17] and methylation patterns of iPSCs compared to embryonic stem cells (ESCs) [8, 18-19]. In addition, key components of pluripotency regulatory networks are being defined (*e.g.,* Oct4 [20], p53 [21], miRNA-145 [22]), and the cellular and molecular aspects of tissue regeneration are being elucidated [23, 24].

Many laboratories now utilize powerful functional genomics approaches to dissect the systems-level processes underlying stem cell biology. Such high-throughput strategies, including conventional microarrays, SNP [9] and miRNA [6] arrays, ChIP-on-chip [4], CHARM methylation profiling [8], and massively parallel sequencing [7] yield very large data sets that are generally deposited with online repositories such as the Gene Expression Omnibus (GEO) (http://www.ncbi.nlm.nih.gov/geo/) or the ArrayExpress archive (http://www.ebi.ac.uk/microarray-as/ae/). Primarily consisting of whole-genome microarray outputs, these pluripotent stem cell data sets have grown in number about three-fold every two years, from five GEO data sets in 2005 to fifty in 2009. Importantly, archived data sets can be reanalyzed to gain new insights into systems-level stem cell biology (*e.g.,* [14, 16-18]), and thus represent a valuable resource for the stem cell community.

Without the analytical power of computational and statistical methods, large genomics data sets are virtually impossible to interpret and thus of limited utility. Unsupervised clustering is one widely used strategy applied to large data sets for identifying groups of similar patterns, such as, for example co-regulated transcripts (for review, see [25]). Although many kinds of unsupervised clustering methods are available, commonly used approaches, like K-Means and Hierarchical clustering, have major limitations for identifying natural data clusters. For instance, K-Means is restricted to symmetrical clusters, is unable to detect outlier data points, and requires prior knowledge of cluster number or use of an external cluster number prediction method [26]. Hierarchical clustering methods, such as those implemented in Eisen's widely used Cluster 3.0 software [27], are also unable to identify the number of clusters [28], make irreversible local decisions that can decrease cluster quality [29], and are inefficient on large genomics data sets [28]. These methods, and many others (see [30]), are less than ideal for researchers seeking to identify and study biologically meaningful patterns from the large amounts of data generated by high-throughput technologies.

To address limitations with the most commonly used clustering strategies, we recently developed and validated a new computational method, called AutoSOME, capable of automatic clustering of large, high-dimensional data sets, such as whole-genome microarray expression data, without prior knowledge of data structure or cluster number [31]. The AutoSOME method is based on a serial application of well-established techniques from different fields, including machine learning, cartography, and graph theory. As demonstrated by the finding of a large protein-protein interaction (PPI) network up-regulated in pluripotent stem cells [31] and by the finding that pluripotent stem cells exhibit lab-specific gene expression signatures [17], AutoSOME provides a valuable approach for analyzing the genetic relationships among different cell lines from large genomics data sets. The AutoSOME method is implemented in Java and packaged within a Graphical User Interface (GUI) to accommodate end-users with diverse backgrounds using diverse computer operating systems (http://jimcooperlab.mcdb.ucsb.edu/autosome).

Here we propose a standard protocol for whole-genome expression analyses based on AutoSOME clustering, and present a primer illustrating the use of the AutoSOME GUI for exploring stem cell gene expression data. Although the protocols in this chapter utilize specific publicly available microarray data sets, the workflow is sufficiently general for use with any number of diverse in-house or publicly available gene expression data. Key steps preceding cluster analysis are described first, including how to import, filter, and normalize microarray gene expression data using built-in GUI functions. Major cluster parameters are subsequently reviewed followed by two detailed examples that demonstrate how to perform an AutoSOME cluster analysis on stem cell microarray data.

Importing Gene Expression Data

AutoSOME accepts two major input file formats. The first input format is a table of numerical values, as shown in Table **1**, with one column of unique gene labels (left column) and one row of array labels (top row). All data entries need to be tab-, comma-, or space-delimited. The second major input format, called a Gene Expression Omnibus Series Matrix File, available online at http://www.ncbi.nlm.nih.gov/geo/, consists of normalized gene expression data generated from a microarray experiment deposited in the GEO archive. AutoSOME can read expression data from a series matrix file, enabling rapid microarray re-analysis. Once imported, gene expression data are represented as a matrix composed of *n* data rows, or gene probes, and *m* data columns, or arrays.

Table 1: Basic Input Format.

Probe	hESC-1	hESC-2	hESC-3	hESC-4	iPSC-1	iPSC-2	iPSC-3
212853_at	8.22	8.29	7.69	8.22	10.13	10.26	10.22
212854_x_at	8.52	8.71	8.04	9.00	8.88	8.97	9.08
212855_at	10.64	10.41	10.60	11.04	12.09	11.91	12.05

Microarray Data Preprocessing

To reduce noise and increase cluster quality, several preprocessing procedures for gene probe filtration and microarray normalization are available in the AutoSOME GUI. The procedures described below are tailored for *intensity* microarray data. For *two-colored* expression data, different normalization steps will be required.

Data Filtration

Gene probe filtration is a common preprocessing step for whole-genome microarray data cluster analysis. By removing gene probes corresponding to transcripts with low background-level expression or low variance across experiments, filtration can decrease algorithm running time and increase the overall signal-to-noise ratio. Filtration options currently available in the AutoSOME GUI are the removal of transcripts with a fold change less than X and/or the removal of transcripts with a mean expression value below some threshold Y.

Normalization

Normalizing microarray data is an *essential* preprocessing step to remove technical bias and mitigate the impact of outlier data points on cluster identification. Unfortunately, the most appropriate normalization protocol is not always straightforward (especially to end-users without knowledge of the various normalization techniques commonly employed). The AutoSOME GUI implements several major normalization methods for data clustering, each of which is described below with recommendations for proper usage (G=recommended for clustering genes, or data rows; A=recommended for clustering arrays, or data columns). All normalization procedures are conducted from top to bottom in the order listed. For technical descriptions, see the manual [27] at http://rana.lbl.gov/manuals/ClusterTreeView.pdf.

Log$_2$ Scaling (G, A)

By amplifying small-scale changes in gene expression, log$_2$ scaling prevents transcripts with low levels of expression from being overshadowed by more highly expressed transcripts. Since AutoSOME works best in the log space, this data adjustment strategy should be used whenever expression values span several orders of magnitude over the entire microarray data set (*e.g.,* 0.3-20, 000). Importantly, unlike data normalization methods, log$_2$ transformation is a scaling procedure that is completely reversible.

Unit Variance (G, A)

Based on the assumption that all arrays have a normal distribution, this technique standardizes expression values within each array to a mean of 0 and standard deviation of 1. When there is no need to treat arrays differently, we strongly recommend using unit variance normalization (even after raw microarray data have been pre-normalized by RMA, MAS5, *etc.*).

Median Centering (G)

Median centering sets the median of each row and/or column equal to zero. In the context of microarrays, this procedure centers the expression pattern, or "waveform", of each gene (or array) so that expression patterns can be isolated and compared without being affected by differences in transcript abundance. We highly recommend applying median-centering to rows in all cases where AutoSOME will be used to identify genes with similar co-expression signatures.

Sum of Squares=1 (G)

Sum of Squares=1 normalization yields substantial data smoothing by setting the sum of squares (x^2) of all expression values equal to 1 for each gene over all arrays and/or each array over all genes. The method has a significant effect on cluster identification, and tends to identify large coherent clusters trailing off into genes with minimal differential expression (background noise can be removed by filtering the data prior to clustering or by using the AutoSOME confidence filter after clustering, see *confidence filter* in the online AutoSOME manual at http://jimcooperlab.mcdb.ucsb.edu/autosome/files/AutoSOME_Manual.pdf. For an example of clusters identified using sum of squares normalization, see the heat map presented in Fig. **6**A of [31]. We recommend applying sum of squares normalization to both rows and columns whenever AutoSOME is used to cluster a microarray data set that has been previously filtered to remove genes with minimal variance (see Data Filtration above).

For additional data adjustment options, one can use Microsoft Excel or the Cluster 3.0 software tool [17] before importing the data into AutoSOME.

MATERIALS

For the protocols described herein, the following software and data sets are needed:

AutoSOME and Java

Download AutoSOME from http://jimcooperlab.mcdb.ucsb.edu/autosome/download.jsp. Since AutoSOME is implemented in Java, in principle, it can be run using any operating system with Java Standard Edition (Java SE) 1.6+, available from http://www.oracle.com/technetwork/java/index.html. In general, we recommend using a computer with at least 1.6GB RAM and at least a dual-core CPU. Of course, the more memory and dedicated cores, the better the performance. Microarray data sets like the Affymetrix HG-U133plus2 chipset (>54k probes) with several samples can be run with 1.6GB RAM (maximum

RAM that can be allocated for 32-bit Java systems), however, large data sets with many arrays (*e.g.*, >200) will benefit from the additional RAM made possible by systems with the 64-bit Java Runtime Environment (up to ~30GB RAM).

Cytoscape

Cytoscape [32], a network visualization tool, should be installed on your computer to run the second protocol (download from http://cytoscape.org). We used Cytoscape 2.6.0; if any aspects of the protocol are not reproducible using the current version of Cytoscape, one can obtain version 2.6.0 from the Cytoscape download page.

Microarray Data

The protocols in this book chapter make use of two publicly available GEO data sets, GSE22651 (http://www.ncbi.nlm.nih.gov/projects/geo/query/acc.cgi?acc=GSE22651) and GSE19164 (http://www.ncbi.nlm.nih.gov/projects/geo/query/acc.cgi?acc=GSE19164). The hyperlink for the Series Matrix File corresponding to each data set is located at the bottom of the GEO data set page under *Download family*. Unzip each Series Matrix File using a decompression tool that can handle the '.gz' format (*e.g.*, 'WinRaR', available at http://www.rarlab.com) and save it to your hard drive.

RUNNING AUTOSOME

AutoSOME can be launched from the AutoSOME website *via* Java Web Start, or by downloading the executable. To use the downloadable version (http://jimcooperlab.mcdb.ucsb.edu/autosome/download.jsp), unzip all contents to the same directory. If you are using Windows, you can run AutoSOME by double-clicking on one of the batch files that come with the download (*e.g.*, runautosome-1.6GB-win32.bat). Otherwise, navigate to the directory on your system where AutoSOME is installed, and run the following command:

java -Xmx1600m -Xms1600m -jar autosome_vXXXXXX.jar

(XXXXXX = MMDDYY represents the date of compilation; *e.g.*, 122911). The -Xmx, -Xms arguments allocate additional memory (in megabytes) to AutoSOME for running large datasets. All, or most, available memory should be allocated. In operating systems running 32-bit Java, the maximum amount of memory that can be allocated to AutoSOME is about ~1.6 GB, while an operating system running 64-bit Java can allocate up to ~30 GB of memory. To see which Java version is installed on your computer, launch a terminal window and type "java -version".

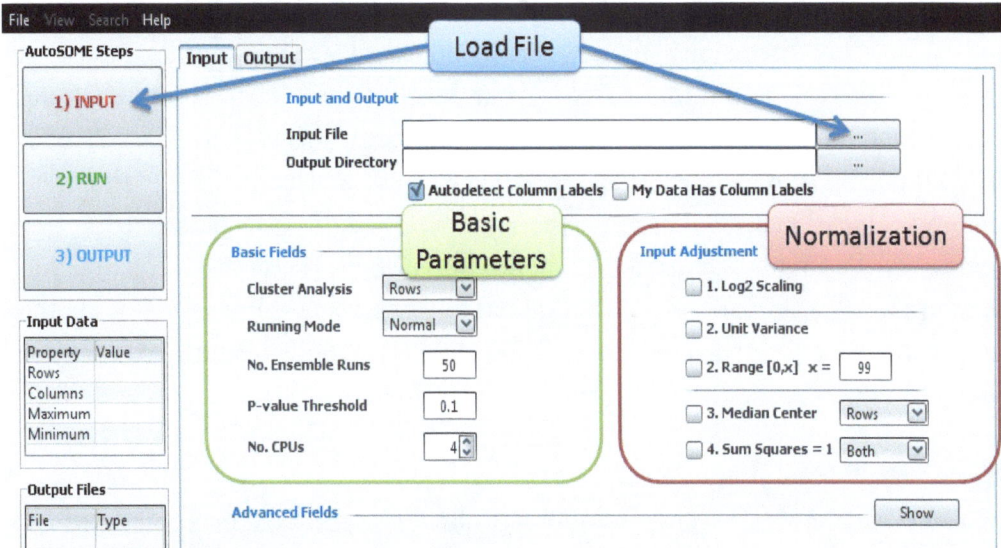

Figure 1: Layout of AutoSOME GUI main window.

BASIC PARAMETERS

A screenshot of the AutoSOME GUI layout is presented in Fig. (**1**). We recommend reviewing the basic AutoSOME parameters, described in this section, before running the protocols in this chapter. These parameters control key aspects of the AutoSOME cluster analysis, including whether AutoSOME will cluster genes and/or array experiments, how long AutoSOME will take to run, and the statistical significance, or granularity, of the cluster output. This section reviews the 'Basic Fields' parameters: Cluster Analysis, Running Mode, No. Ensemble Runs, P-value Threshold, and No. CPUs (see Basic Parameters in Fig. **1**).

Cluster Analysis

Use the Cluster Analysis combo box to toggle among clustering rows (genes), columns (arrays), or both (genes followed by arrays).

Running Mode

Use the Running Mode combo box to select among 'Precision', 'Normal', or 'Speed' modes of operation. Each mode specifies different parameters for two major components of the AutoSOME method, Self-Organizing Map (SOM, see [33]) and density-equalization. Greater training of the SOM and greater resolution of density-equalization can lead to more accurate delineation of cluster boundaries. The 'Precision' mode takes longest (SOM=2X1000 iterations, density-equalization resolution=64X64), but has the best chance of resolving difficult cluster borders. On the other hand, 'Speed' (SOM iterations=2X250, density-equalization resolution=16X16) is very rapid, and is useful for first-pass exploratory cluster analysis. In our experiments, a compromise between the two extremes, 'Normal' (SOM iterations=2X500, density-equalization resolution=32X32), generally yields comparable results to 'Precision' with the benefit of increased speed. Depending upon desired expediency, we recommend selection of either 'Normal' or 'Precision' for final clustering results.

Ensemble Runs

AutoSOME stochastically samples a large cluster space and makes use of an ensemble averaging procedure to stabilize the cluster output. As demonstrated in Newman and Cooper, 2010a, increasing ensemble iterations can dramatically reduce output variance and increase cluster quality. Additional ensemble stability tests indicate that gene co-expression clusters in noisy whole-genome microarray data exhibit the greatest gain in cluster stability by 50 ensemble iterations (data not shown). Co-expression clusters continue to gradually stabilize with increasing iterations past 50. While 50 ensemble iterations (default) is enough to investigate the cluster structure of most data sets, we recommend using fewer ensemble iterations (*e.g.,* 10-20) for a first-pass exploratory analysis and using 100-500 iterations for a final clustering.

P-Value Threshold

A critical step of the AutoSOME method involves partitioning a graph containing all input data points into a set of data clusters. The p-value threshold allows the data graph to be cut into statistically significant clusters based on a simulated null hypothesis of random data points. The smaller the p-value the tighter (and smaller) the resulting clusters. A default threshold of ≤ 0.1 has been extensively benchmarked to yield consistently good accuracy on a wide variety of clustering problems (see [31]). Lower the p-value threshold for increasingly challenging datasets or increasingly fine-grained clusters.

No. CPUs

Due to the ensemble averaging step, the running time of AutoSOME will reduce linearly with respect to an increasing number of dedicated CPUs, and thus, all available CPU cores are allocated by default.

PROTOCOLS

This section consists of two general protocols that illustrate the use of AutoSOME for large-scale exploration of gene expression signatures. Both protocols make use of the publicly available software and data sets described in the Materials section (above). The first protocol demonstrates how to use AutoSOME to identify both gene co-expression modules and transcriptome clusters from publicly available microarray data. The second protocol shows how to use Cytoscape [32] to visualize stem cell transcriptome variation using an intuitive two-dimensional network schematic called a "fuzzy cluster network" (*e.g.,* see Fig. **3** in [17]).

PROTOCOL 1: AUTOSOME CO-EXPRESSION AND TRANSCRIPTOME CLUSTERING

Here we show how to cluster both transcripts and transcriptomes in a single efficient step with the AutoSOME GUI. By executing this protocol, you will also be introduced to important output features of the AutoSOME GUI, including displaying and adjusting heat maps, and saving publication quality figures of the cluster output. Due to the inherent noise in microarray data sets and the stochastic component of AutoSOME, your cluster results may vary slightly (and only slightly) from those presented here.

1) Launch AutoSOME and press the large 'INPUT' button (see Fig. **1**). A file browser will appear. Select the 'Gene Expression Omnibus Series Matrix File' checkbox located in the data format box over the browsing window (otherwise there will be an input error). Browse to

where you saved the GSE22651 text file (should have been saved as 'GSE22651_series_matrix.txt', see Materials), select it, and press the 'Open' button.

2) The number of arrays (=65) and gene probes (=48, 786), along with maximum and minimum data values, will be shown. You are asked whether you would like to filter your data. Press 'Yes'. A 'Filter Data' window will open.

3) Since the expression values span a range of 26 to 27, 446, the data have not been $\log_2$ scaled. We can leave the corresponding checkbox deselected. To reduce the computational load and generate cleaner clusters, remove gene probes with a fold change (maximum over minimum value) less than 4. Press 'Apply' to preview the filtered data results. The filtered data set has 14, 990 rows (gene probes). Press 'Accept'.

4) In the main GUI window, expand 'Basic Fields' by pressing 'Show'. Since AutoSOME will be used to cluster both filtered transcripts and transcriptomes, select 'Both' from the 'Cluster Analysis' combo box. 'Basic Fields' will expand to 'Basic Fields (Rows)' and 'Basic Fields (Columns)'. Under 'Basic Fields (Rows)', set 'Running Mode'=Normal, 'No. Ensemble Runs' to 100, 'P-value Threshold' =0.1, and underneath in 'Basic Fields (Columns)', set 'No. Ensemble Runs' to 200 and 'P-value Threshold' =0.1.

5) Expand 'Input Adjustment'. Under 'Input Adjustment (Rows)', select '$\log_2$ Scaling', 'Unit Variance', 'Median Centering' of 'Rows', and 'Sum Squares = 1' 'Both'. Under 'Input Adjustment (Columns)', select '$\log_2$ Scaling' and 'Unit Variance'.

6) **Note**: AutoSOME can also be used to cluster unfiltered microarray data (and even larger data sets). In this case, 'Sum of Squares=1' normalization is **not** recommended.

7) Press the large 'RUN' button. Progress and elapsed time are shown in the 'Run Progress' box located in the lower-left region of the main

GUI window. Note that running time will vary depending upon the number of dedicated CPUs and CPU clock speed. (For this data set a typical run time of ~10 minutes should be expected using a 3GHz dual CPU computer).

8) **Note**: if AutoSOME does not finish, there may not be enough memory available. If you have more RAM on your computer, simply allocate additional RAM to AutoSOME at start-up (1.6GB is sufficient; see Running AutoSOME). There is also an option to write intermediate ensemble runs to disk (go to 'Advanced Fields'>Memory from the main window of the GUI). If selected, a temporary folder will be created in the current working directory.

9) Once clustering is finished, AutoSOME will write output files to disk (see Table **2**), and the main window will be redirected to the 'Output' tab. AutoSOME will identify 3 array (or transcriptome) clusters and approximately 42 gene (or co-expression) clusters. Co-expression clusters are displayed as a list in the 'Cluster Output' tree with the number of transcripts in each cluster shown in parentheses. Select 'cluster 1' with your mouse. By default, a table of all gene probe labels in cluster 1 (along with cluster confidence values, see Newman and Cooper 2010a) will be displayed.

Table 2: AutoSOME Output Files. By default, all output files will be written to the parent directory of your input file. ('MyInput' = name of input file, 'X' = the number of ensemble runs, 'Y' = the p-value, and 'Z' = 'rows' or 'columns' depending upon what whether genes or arrays were clustered, respectively (*e.g.,* AutoSOME_GSE22651_E100_Pval0.1_rows.txt)).

Cluster Rows or Columns	File Name	Description	Open in GUI
Either	AutoSOME_MyInput_EX_PvalY_Z_summary.html	List of all AutoSOME parameters, and cluster summary table for either row or column clustering	No
Either	AutoSOME_MyInput_EX_PvalY_Z.html	HTML version of all clusters and confidence values for either row or column clustering	No
Either	AutoSOME_MyInput_EX_PvalY_Z.txt	Text file of all AutoSOME clusters and confidence values for either row or column clustering (stores original data prior to normalization)	Yes

Table 2: cont...

Both	AutoSOME_MyInput_PvalY_rows_columns.txt	Text file of all AutoSOME clusters and confidence values for both row and column clustering (stores original data prior to normalization)	Yes
Columns	AutoSOME_MyInput_EX_PvalY_Edges.txt	Fuzzy cluster network edges and edge weights for use with Cytoscape [Shannon *et al.*, 2003]	No
Columns	AutoSOME_MyInput_EX_PvalY_Nodes.txt	Fuzzy cluster network nodes for use with Cytoscape [Shannon *et al.*, 2003]	No
Columns	AutoSOME_MyInput_EX_PvalY_Matrix.txt	Fuzzy cluster network edge weights in matrix form (can be hierarchically clustered using Cluster 3.0 [27] and visualized in Java TreeView [34])	No

10) While cluster 1 is still selected, select 'View'>'heatmap'>'green red' from the dropdown menu. A traditional heat map will be displayed. A scale-bar above the heat map shows the maximum and minimum normalized expression values. A red-blue color bar on the left side of the main heat map indicates the confidence of each gene probe for its cluster (Fig. **2A**, blue=100% confidence, red=0%, see Fig. **2** in [31]). Since we clustered arrays as well as gene probes, white vertical bars are shown separating different array clusters (Fig. **2A**).

11) To display heat maps of more than one gene co-expression cluster simultaneously, hold Shift or Ctrl to select multiple clusters from the cluster list using your mouse. Let's select clusters 1-10 (hold Shift). Since the heat map is too large to see all clusters without scrolling, go to 'View'>'fit to screen'. Now all 10 co-expression clusters are visible separated by white bars (see Fig. **2A**). (If some clusters extend beyond the display after using the fit screen function, the clusters at the bottom may be viewed by using the scroll bar, changing the image dimensions using the 'image settings' window (see below), or using the mouse scroll wheel.)

12) To more clearly visualize expression differences in the heat map, we can change the normalization settings and heat map contrast. Go to 'View'>'settings'>'image settings' to launch a new window. The 'Image Settings' window will appear showing a wide variety of adjustable display settings (Fig. **2B**). To use the heat map to accurately display

expression differences, we need show the data prior to normalization. To do this, select 'Display Original Data'. Since the raw expression data span a very large range, select Log_2 scaling. Next, select 'Median Center Rows' to center all gene probes. Finally, let us adjust the heat map contrast. You can slide the 'Heat Map Contrast' bar to the left (*e.g.,* to 0.2) and let go, or for more precise contrast adjustment, select 'Manually adjust range for contrast' and input explicit minimum and maximum values, such as, for example, -2 to 2. For the typed-in manual adjustments to take effect it is then necessary select the update button. Note that the color-bar in the heat map updates to reflect the new value range. Finally, switch to the 'Display Options' tab and select 'Hide Heat Map Row Labels' to remove gene probe labels from the right side of the heat map (these cannot be seen at the current resolution). The heat map should look similar to the one shown in Fig. **2B**.

13) To save a high-resolution image of this heat map, increase the 'Zoom Factor' bar to 50 (zoom factor is the width in pixels of each column in the heat map). Let's leave the 'Adjust Height' bar the same since it was determined by the 'fit to screen' command (the height of each row in the heat map is determined by multiplying this number by the zoom factor). Press the 'Save' button at the bottom of the 'Image Settings' window to write the heat map to disk (Portable Network Graphics (PNG) format only). Although the image will overflow the display window, the entire image will be saved to file. In addition to being able save the heat map images created in the output window, AutoSOME automatically saves additional output files (see Table **2**).

Anticipated Results

The filtered GSE22651 data set has two large gene co-expression clusters that distinguish pluripotent stem cell lines from somatic cell lines (see Fig **2C**). There are also three filtered transcriptome clusters, an undifferentiated pluripotent stem cell cluster and two somatic cell line clusters: a cluster of fibroblasts, mesenchymal cells, keratinocytes, and human umbilical vein endothelial chord cell lines, and a smaller cluster composed of transcriptomes representing lung, adipose, bladder, and ureter tissue samples.

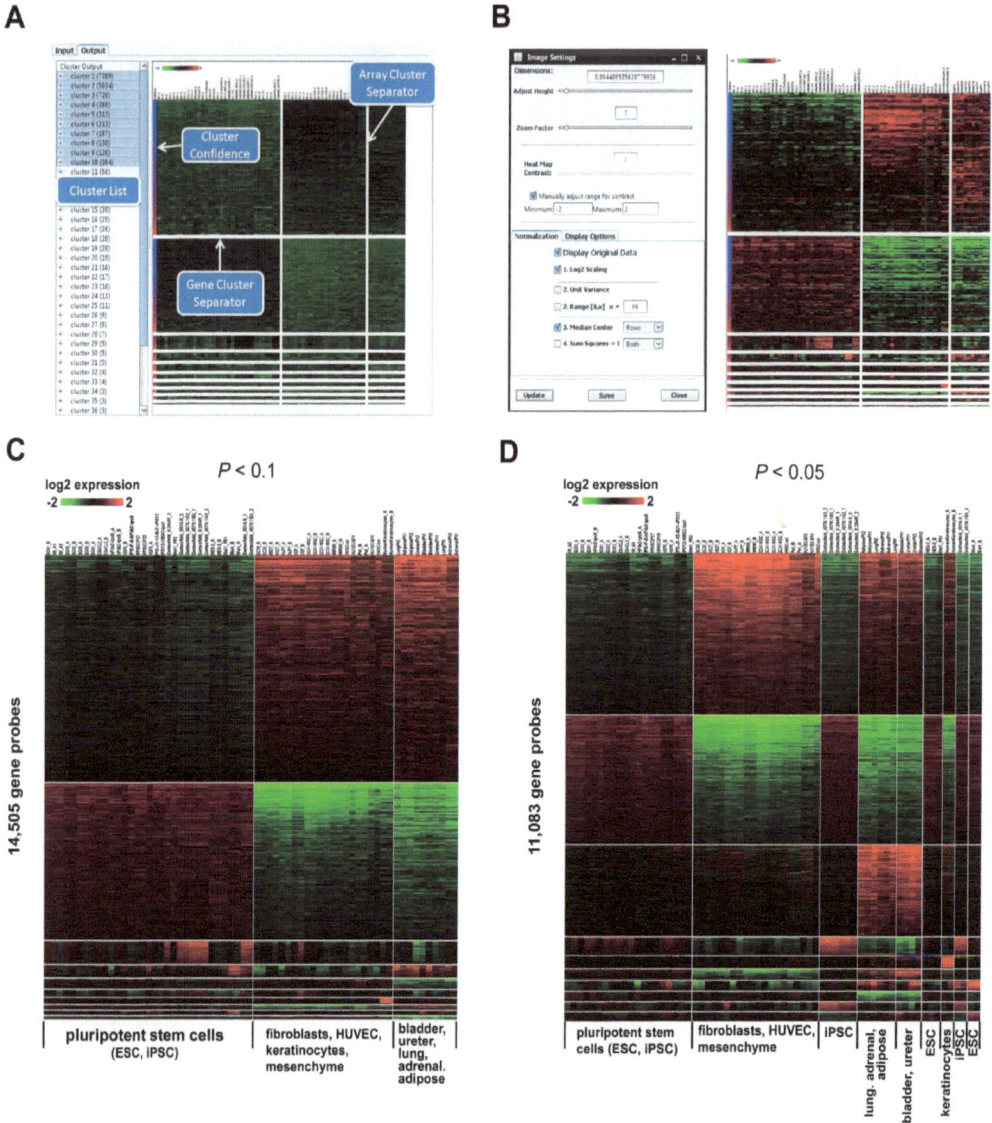

Figure 2: AutoSOME co-expression and transcriptome clustering, related to Protocol 1. (**A**) Cluster list and heat map output, (**B**) Image settings window and renormalized cluster heat map, (**C**) Final heat map for 10 largest gene co-expression clusters at $P < 0.1$ with cellular phenotypes corresponding to each transcriptome cluster shown underneath, (**D**) Final heat map for 10 largest gene co-expression clusters at $P < 0.05$ with cellular phenotypes corresponding to each transcriptome cluster shown underneath, . To render heat maps shown in panels (**C**) and (**D**), follow protocol 1 through step 10 using the appropriate p-value, then reorder each cluster by decreasing variance: go to the 'Display Options' tab in the Image Settings window (panel B) and select 'Sort by Decreasing Variance' (you may need to press 'Update' if the heat map fails to refresh), and finally, follow step 11.

Next Step

Try repeating this protocol using a p-value threshold of 0.05 for both rows and columns. You will notice that AutoSOME identifies tighter clusters, including a distinct keratinocyte transcriptome cluster, and a bladder and ureter transcriptome cluster. In addition, several tighter co-expression clusters are now resolved. Heat maps displaying clusters obtained with p-value thresholds 0.1 and 0.05 are shown in Figs. (**2C**) and (**2D**), respectively.

PROTOCOL 2: EXPLORING STEM CELL TRANSCRIPTOME VARIATION USING A FUZZY CLUSTER NETWORK

A powerful application of AutoSOME clustering, in addition to identifying discrete clusters of co-expressed genes (*e.g.,* Fig **2C**), is identifying data points with fractional membership to one or more clusters. Such "fuzzy clusters" are a natural way of representing the inherent noise in gene expression data, and importantly, when visualized as a network diagram, provide an intuitive schematic for displaying the relationships among the transcriptome clusters identified by AutoSOME. In the first part of this protocol, AutoSOME is used to cluster the transcriptomes of several pluripotent stem cell lines, including iPSCs generated from three different combinations of reprogramming factors [12]. Using the cluster results from part one, the second part of this protocol details how to create a fuzzy cluster network. Before proceeding, it would be useful to become familiar with the transcriptome clustering strategy utilized by AutoSOME, which involves the construction of a distance matrix of transcriptome profiles.

Transcriptome Clustering using a Distance Matrix

The number of microarray gene probes n is usually much greater than the number of arrays m. To decrease the computational load for clustering transcriptomes, AutoSOME does not directly cluster transcriptome profiles, but instead clusters a matrix representing pair-wise similarities of all transcriptomes. (This amounts to performing an All-against-All comparison of m array expression vectors to generate a similarity matrix of size m by m used for clustering.) Three common distance metrics (Euclidean, Pearson's, and uncentered correlation) for calculating transcriptome *similarity* are provided as a user-adjustable parameter. To access the 'Distance Metric' combo box, expand 'Advanced Fields' in the GUI main window and go to 'Fuzzy Cluster Networks'. Euclidean distance is selected by default. AutoSOME also implements Pearson's correlation and uncentered correlation metrics, both of which have a maximum of 1 (completely correlated) and minimum of -1 (inversely correlated). Unlike uncentered correlation, Pearson's correlation is insensitive to amplitude shifts, meaning that two transcriptomes with similar expression patterns but different amplitudes can still be highly correlated using Pearson's method. For an excellent review of these three distance metrics, see [25].

PROTOCOL 2 PART 1: AUTOSOME TRANSCRIPTOME CLUSTERING

1) Launch AutoSOME and press the large 'INPUT' button (see Fig. **1**). A file browser will appear. Select the 'Gene Expression Omnibus Series Matrix File' checkbox located in the data format box over the browsing window (otherwise there will be an input error). Browse to where you saved the GSE19164 text file (should be saved as 'GSE19164_series_matrix.txt', see Materials), select it, and press the *Open* button.

2) You will be asked whether you want to filter the data. Press 'No'.

3) In the main GUI window, expand 'Basic Fields' by pressing 'Show'. Since we are going to cluster cellular transcriptomes, select 'columns' from the 'Cluster Analysis' combo box. Set 'Running Mode'=Normal, 'No. Ensemble Runs' to 500, and 'P-value Threshold' =0.1.

4) Expand 'Input Adjustment'. Since these expression data have not been $\log_2$ scaled, select '$\log_2$ Scaling'. Also, select 'Unit Variance'.

5) At this point, the distance matrix used for column clustering could be changed by expanding 'Advanced Fields' and picking another metric from the 'Distance Metric' combo box. We will use the default setting, Euclidean distance.

6) Press the large 'RUN' button.

7) After clustering is finished, output files will be written to disk (see Table **2**), and the main window will be redirected to the 'Output' tab. Select all clusters in the cluster list using your mouse (hold Shift), and

select 'View'>'heatmap'>'rainbow' from the dropdown menu. The display should look like Fig. (**3**). The heat map shows the clustering of the Euclidean distance matrix of all arrays in the GSE19164 data set (see Choice of Distance Metric). Each transcriptome cluster is separated by a white horizontal bar. The Euclidean distance between any given pair of cellular transcriptomes can be inferred from the heat map by using the color bar. A distance of zero means that the two cell lines are identical by Euclidean distance (colored blue).

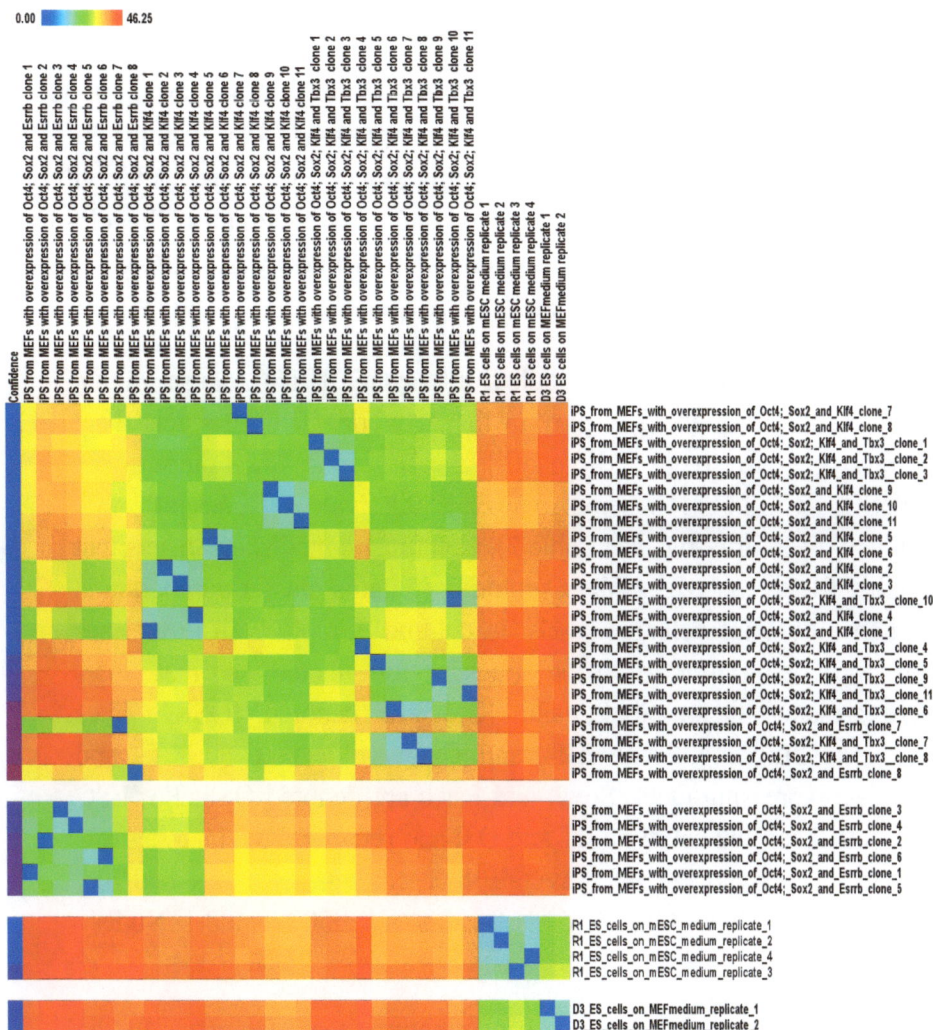

Figure 3: Clustered Euclidean distance matrix of iPSC and ESC cell lines, related to Protocol 2 Part 1.

Anticipated Results

Four transcriptome clusters are identified, as shown in Fig. (**3**). Notice that ESCs and iPSCs cluster separately, and each cell type also has two distinct sub-clusters. To explore relationships among each cluster, along with individual transcriptomes, proceed to part two (below).

Protocol 2 Part 2: Fuzzy Cluster Network

The fuzzy cluster network has three major components: nodes, edges, and clusters. A node represents a particular transcriptome, usually labeled by cellular phenotype or time point. An edge is defined as a link between two transcriptomes (nodes). The edge weight represents the fraction of times over all ensemble runs that the pair of transcriptomes clustered together, minus 0.5. For example, if two transcriptomes clustered together 80% of the time, or 0.8, the edge value will be 0.3. The full range of edge weights is thus -0.5 to 0.5. Finally, the clusters represent the discrete clusters of transcriptomes identified by AutoSOME (*e.g.,* see Fig. **3**). For further details, see [31].

1)   First, let's modify the cell line labels from the GSE19164 data set so that they are short enough to display in the network. Using a spreadsheet editing program (*e.g.,* Microsoft Excel), open the output file 'AutoSOME_GSE19164_series_matrix_E500_Pval0.1_Nodes.txt' (for details of this output file, see Table **2**). Column 1 contains identifiers that link this file to the edges file. Column 2 contains cluster numbers and Column 3 contains cell line labels. In column 3, replace all iPSC lines with the first letter of each transcription factor so that lines overexpressing Oct4, Sox2, and Klf4 are denoted 'OSK', lines overexpressing Oct4, Sox2, Klf4, and Tbx3 are denoted 'OSKT', and lines overexpressing Oct4, Sox2, and Esrrb are denoted OSE. For ESC lines, let's keep the cell type R1 and D3, and remove the remaining text to yield R1_ES and D3_ES. Save all three columns as a new text file, *e.g.,* 'AutoSOME_GSE19164_series_matrix_E500_Pval0.1_Nodes_reformat.txt'.

2)   Launch Cytoscape (for download information, see Materials). In the 'File' dropdown menu, select the option to import a network from a

table (Fig. **4A**). Go to 'Select File (s)' to select the AutoSOME output file containing all edges called 'AutoSOME_GSE19164_series_matrix_E500_Pval0.1_Edges.txt' (for details of this output file, see Table **2**). Press 'Open'. Set 'Source Interaction' to 'Column 1' and 'Target Interaction' to 'Column 2'. Finally, press to activate Column 3 in the data Preview window (it will turn blue). Choose 'Import', and finally, click 'Close'. A raw network will show up as a grid. To render a network with detailed graphics, go to the dropdown menu and select 'View'→'Show Graphics Details'.

3) In the 'File' dropdown menu, select the option to import attributes from a table. Go to 'Select File (s)' and select the AutoSOME output file containing all nodes and modified cell line labels, 'AutoSOME_GSE19164_series_matrix_E500_Pval0.1_Nodes_reform at.txt'. Select 'Open' and then 'Import'.

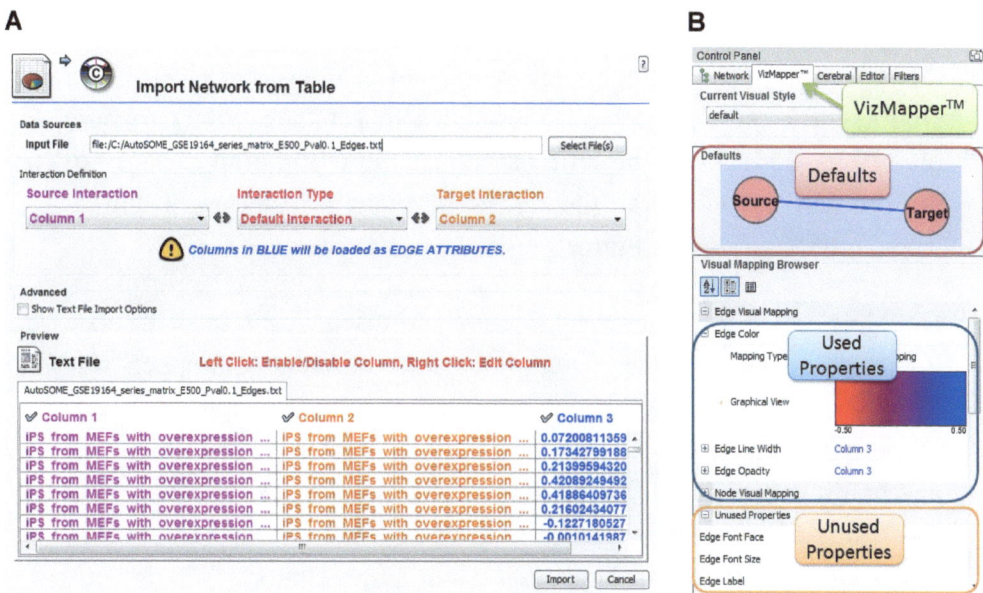

Figure 4: Cytoscape screenshots. (**A**) Import edges into network. (**B**) Major features of the Control Panel.

4) Adjust global properties of the network: In the Control Panel, choose the VizMapper™ tab (Fig. **4B**). Click in the 'Defaults' window (Fig.

4B, displaying a source and target pair with a blue background). A new window will appear. Choose the 'Global' tab in the bottom right. Change the background color to white. Go back to the 'Node' tab and change the NODE_BORDER_COLOR property to black and increase the 'NODE_LINE_WIDTH' to 2. Change NODE_FONT_SIZE to 8. Press 'Apply'. The network should look like Fig. (**5A**).

5) Using the VizMapper™ tab: Next to 'Node Label', press 'ID' and choose 'Column 3'. All nodes are then relabeled according to the original data labels. Minimize the 'Node Label' property by choosing the minus icon. The network will now look like Fig. (**5B**).

6) Under 'Unused Properties' (Fig. **4B**): double-click 'Edge Color' (top of list). Choose 'Column 3' as a value. Then, choose 'Continuous Mapper' for 'Mapping Type'. Click the black-to-white gradient next to 'Graphical View' to launch a Gradient Editor. There are two fixed triangles, one on each end, and two adjustable triangles. Double-click the two leftmost triangles to set their colors to pure red (255, 0, 0). Double-click the two rightmost triangles to set their colors to pure blue (0, 0, 255). Drag the leftmost adjustable triangle all the way to the left and likewise drag the rightmost triangle to the right until it stops. Exit 'Continuous Editor'.

7) Double-click 'Edge Line Width' under 'Unused Properties'. Choose 'Column 3' as a value. Then, choose 'Continuous Mapper' for 'Mapping Type'. Click on the graph next to the 'Graphical View' property to launch the 'Continuous Editor'. Adjust the minimum and maximum values denoted by red squares (double-click on squares for precision, otherwise slide squares up or down, *e.g.,* set minimum to 0.5 and maximum to 10). Then, exit 'Continuous Editor'.

8) Double-click it 'Edge Opacity' under 'Unused Properties'. Choose 'Column 3' as a value. Then, choose 'Continuous Mapper' for 'Mapping Type'. Click on the graph next to the 'Graphical View' property to launch the 'Continuous Editor'. Adjust the minimum and

maximum values denoted by red squares (double-click on squares for precision, otherwise slide squares up or down, *e.g.,* set minimum to 0.5 and maximum to 150). Then, exit 'Continuous Editor'.

9) Finally, double-click 'Node Color' under 'Unused Properties'. Choose 'Column 2' as a value. Then, choose 'Discrete Mapper' for 'Mapping Type'. Right click on 'Discrete Mapping' and go to 'Generate Discrete Values'→'Rainbow 1'. All nodes are then colored according to cluster labels. Adjust colors as desired. The network should now look like Fig. (**5C**).

10) Select 'Layout' in the dropdown menu (top of main window) and choose 'Settings'. Select 'Force-Directed Layout' for 'Layout Algorithm'. Under 'Edge Weight Settings', choose Column 3 from 'The edge attribute that contains the weights', set 'The minimum edge weight to consider' to -0.5, and set 'The maximum edge weight to consider' to 0.5. Click 'Execute Layout' to run the layout algorithm (see, *e.g.,* Fig. **5D**). To increase the repulsion between neighboring nodes (for evenly spaced nodes within a cluster), increase 'Default Node Mass' under 'Algorithm settings'. Different runs of the layout algorithm will yield slightly different results in terms of network rotation and local node placement, while the network topology is generally preserved. Additionally, with different data sets, it is not uncommon for the layout algorithm to generate a network that is logically flawed (*i.e.,* with closely related nodes, as indicated by thick, dark blue edge lines, being distantly removed from each other). When this happens, simply re-execute the layout until a logically consistent network is rendered.

11) Note: Another layout algorithm that allows comparable results is the 'Edge-weighted Spring Embedded' algorithm. Before trying the layout, make sure the 'Edge Weight Settings' are adjusted as above. This layout algorithm can result in more evenly spaced nodes, but is less stable than 'Force-Directed Layout'. Run a few times.

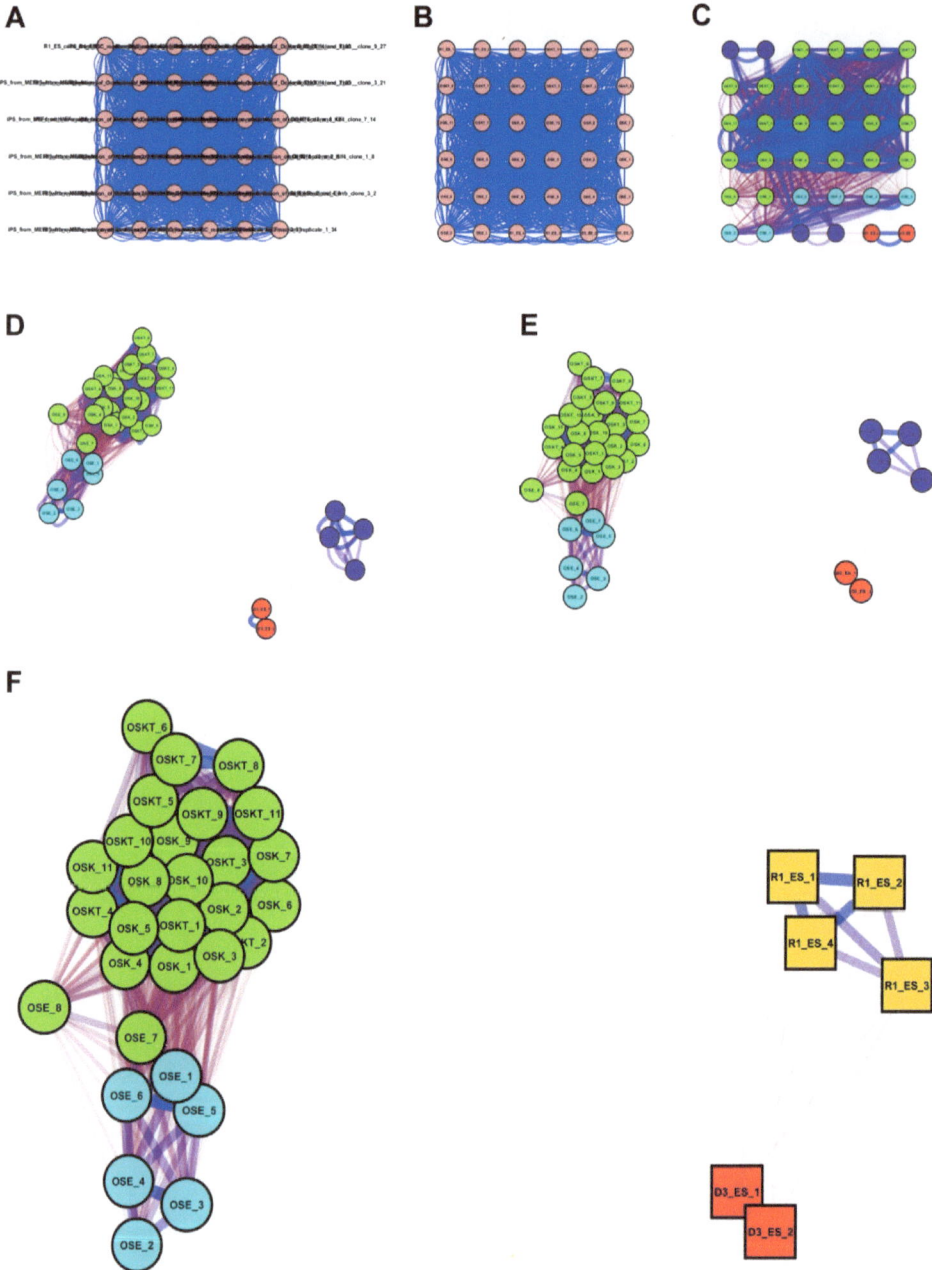

Figure 5: How to render an AutoSOME fuzzy cluster network using Cytoscape, related to Protocol 2 Part 2. (**A**) After importing edges. (**B**) After importing nodes and changing node labels. (**C**) After adjusting node and edge settings. (**D**) After performing layout algorithm. (**E**) After rotating network. (**F**) Final network with iPSCs denoted by circles, ESCs denoted by squares, and color adjustment of R1_ESCs to orange.

12) If desired, rotate (go to 'Select' in the dropdown menu and press 'Rotate') and/or zoom the network until it looks aesthetic (see, *e.g.,* Fig. **5E**). Some nodes may overlap with others. You may be able to manually nudge them into view without substantially altering the network topology (*e.g.,* Fig. **5E**).

13) To output the final network, go to 'File'→'Export'→'Network View as Graphics…'. Then, choose file format and save the image.

Anticipated Results

The final fuzzy cluster network figure is shown in Fig. (**5F**). Note that nodes represent transcriptomes, edge weights represent the fraction of times that each pair of transcriptomes clustered together over all ensemble runs, and differently colored nodes represent discrete clusters from Protocol 2 Part 1 (Fig. **3**). Clearly, iPSCs are more similar to each other than to ESCs, and *vice versa*. Two clusters distinguish iPSCs overexpressing Oct4, Sox2, and Esrrb (OSE) from iPSCs overexpressing Oct4, Sox2, and Klf4 (OSK) or Oct4, Sox2, Klf4, and Tbx3 (OSKT). In addition, OSE lines 7 and 8 bridge both the OSE and OSK/OSKT lines, and OSK and OSKT lines are indistinguishable (at least at $P < 0.1$). These results are comparable to the hierarchical tree shown in Fig. **3** of [12].

CONCLUSIONS

Clustering is the process of partitioning information into useful categories. Although the human brain is endowed with powerful classification tools, our innate faculties for identifying data clusters are challenged by the massive, often high-dimensional, data sets made possible by twenty-first century high-throughput technologies. We developed a new unsupervised clustering method for genomics research, called AutoSOME, to overcome important limitations of common clustering methods, including poor scalability to large data sets, cluster shape restrictions, lack of outlier detection, and most importantly, inability to determine the number of data clusters [31]. In this chapter, we demonstrated how AutoSOME clustering can be applied to stem cell genomics research. Specifically, we presented a primer illustrating how to use the AutoSOME GUI for microarray filtration and normalization, and for "single step" co-expression and transcriptome clustering. In addition, we showed how one can visualize transcriptome variation among stem cell lines by rendering an AutoSOME fuzzy cluster network diagram. Taken together, the workflow proposed in this chapter has utility for studying gene expression signatures in diverse cellular phenotypes and systems, and should have broad application for clustering genomics data generated by diverse microarray platforms and massively parallel sequencers.

CONFLICT OF INTEREST

None declared.

ACKNOWLEDGEMENTS

We thank Dr. Monte Radeke, Dr. Don Anderson, and Dr. Chris Banna for testing the protocols and Dr. Monte Radeke for critically reading this manuscript.

REFERENCES

[1]   Takahashi K, Yamanaka S. Induction of pluripotent stem cells from mouse embryonic and adult fibroblast cultures by defined factors. Cell 2006; 126: 663-676.
[2]   Hayden EC. California stem-cell grants awarded. Nature 2009; 462: 22.
[3]   Wadman M. Most popular cell lines close to approval for US federal funding. Nature 2010; 464: 967.
[4]   Komashko VM, Acevedo LG, Squazzo SL, *et al.* Using ChIP-chip technology to reveal common principles of transcriptional repression in normal and cancer cells. Genome Res 2008; 18: 521-532.
[5]   Müller FJ, Laurent LC, Kostka D, *et al.* Regulatory networks define phenotypic classes of human stem cell lines. Nature 2008; 455: 401-405.
[6]   Wilson KD, Venkatasubrahmanyam S, Jia F, Sun N, Butte AJ, Wu JC. MicroRNA profiling of human-induced pluripotent stem cells. Stem Cells Dev 2009; 18: 749-758.
[7]   Guttman M, Garber M, Levin JZ, *et al.* Ab initio reconstruction of cell type-specific transcriptomes in mouse reveals the conserved multi-exonic structure of lincRNAs. Nat Biotechnol 2010; 28: 503-510.
[8]   Kim K, Doi A, Wen B, *et al.* Epigenetic memory in induced pluripotent stem cells. Nature doi:10.1038/nature09342.   2010   July   19.   Available   from: http://www.nature.com/nature/journal/vnfv/ncurrent/full/nature09342.html
[9]   Närvä E, Autio R, Rahkonen N, *et al.* High-resolution DNA analysis of human embryonic stem cell lines reveals culture-induced copy number changes and loss of heterozygosity. Nat Biotechnol 2010; 28: 371-377.
[10]   Yu J, Hu K, Smuga-Otto K, *et al.* Human induced pluripotent stem cells free of vector and transgene sequences. Science 2009; 324: 797-801.

[11]   Kim D, Kim CH, Moon JI, *et al.* Generation of human induced pluripotent stem cells by direct delivery of reprogramming proteins. Cell Stem Cell 2009; 4: 472-476.

[12]   Han J, Yuan P, Yang H, *et al.* Tbx3 improves the germ-line competency of induced pluripotent stem cells. Nature 2010; 463: 1096-1100.

[13]   Zhao XY, Li W, Lv Z, *et al.* iPS cells produce viable mice through tetraploid complementation. Nature 2009; 461: 86-90.

[14]   Chin MH, Mason MJ, Xie W, *et al.* Induced pluripotent stem cells and embryonic stem cells are distinguished by gene expression signatures. Cell Stem Cell 2009; 5: 111-123.

[15]   Marchetto MCN, Yeo GW, Kainohana O, Marsala M, Gage FH, Muotri AR. Transcriptional signature and memory retention of human-induced pluripotent stem cells. PLoS ONE 2009; 4: e7076.

[16]   Ghosh Z, Wilson DK, Wu Y, Hu S, Quertermous T, Wu JC. Persistent donor cell gene expression among human induced pluripotent stem cells contributes to differences with human embryonic stem cells. PLoS ONE 2010; 5: e8975.

[17]   Newman AM, Cooper JB. Lab-specific gene expression signatures in pluripotent stem cells. Cell Stem Cell 2010; 7: 258-262.

[18]   Guenther MG, Frampton GM, Soldner F, *et al.* Chromatin structure and gene expression programs of human embryonic and induced pluripotent stem cells. Cell Stem Cell 2010; 7: 249-257.

[19]   Polo JM, Liu S, Figueroa ME, *et al.* Cell type of origin influences the molecular and functional properties of mouse induced pluripotent stem cells. Nat Biotechnol 2010; 28: 848-855.

[20]   van den Berg DL, Snoek T, Mullin NP, *et al.* An Oct4-centered protein interaction network in embryonic stem cells. Cell Stem Cell 2010; 6: 369-381.

[21]   Hong H, Takahashi K, Ichisaka T, *et al.* Suppression of induced pluripotent stem cell generation by the p53-p21 pathway. Nature 2009; 460: 1132-1135.

[22]   Xu N, Papagiannakopoulos T, Pan G, Thomson JA, Kosik KS. MicroRNA-145 regulates OCT4, SOX2, and KLF4 and represses pluripotency in human embryonic stem cells. Cell 2009; 137: 647-658.

[23]   Kragl M, Knapp D, Nacu E, *et al.* Cells keep a memory of their tissue origin during axolotl limb regeneration. Nature 2009; 460: 60-65.

[24]   Pajcini KV, Corbel SY, Sage J, Pomerantz JH, Blau HM. Transient inactivation of Rb and ARF yields regenerative cells from postmitotic mammalian muscle. Cell Stem Cell 2010; 7: 198-213.

[25]   D'haeseleer. How does gene expression clustering work? Nat Biotechnol 2005; 23: 1499-1501.

[26]   Xu R, Wunsch II D. Survey of clustering algorithms. IEEE Trans Neural Netw 2005; 16: 645-678.

[27]   Eisen MB, Spellman PT, Brown PO, Botstein D. Cluster analysis and display of genome-wide expression patterns. Proc Natl Acad Sci USA 1998; 95: 14863-14868.

[28]   Giancarlo R, Scaturro D, Utro F. Computational cluster validation for microarray data analysis: experimental assessment of Clest, Consensus Clustering, Figure of Merit, Gap Statistics and Model Explorer. BMC Bioinform 2008; 9: 462.

[29]   De Souto MCP, Costa IG, de Araujo DSA, Ludermir TB, Schliep A. Clustering cancer gene expression data: a comparative study. BMC Bioinform 2008; 9: 497.

[30]   Andropoulos B, An A, Wang X, Shroeder M. A roadmap of clustering algorithms: finding a match for a biomedical application. Briefings Bioinf 2009; 10: 297-314.

[31]   Newman AM, Cooper JB. AutoSOME: a clustering method for identifying gene expression modules without prior knowledge of cluster number. BMC Bioinformatics 2010; 11: 117.

[32]   Kohonen T. The self-organizing map. Proc of the IEEE 1990; 78: 1464-1480.

[33]   Shannon P, Markiel A, Ozier O, *et al.* Cytoscape: a software environment for integrated models of biomolecular interaction networks. Genome Res 2003; 13: 2498-2504.

[34]   Saldanha AJ. Java Treeview-extensible visualization of microarray data. Bioinformatics 2004; 20: 3246-3248.

CHAPTER 4

Image-Enhanced Systems Biology: A Multiscale, Multidimensional Approach to Modeling and Controlling Stem Cell Function

George Plopper[1], Melinda Larsen[2] and Bülent Yener[3,*]

[1]Rensselaer Polytechnic Institute; [2]University at Albany, State University of New York and [3]Rensselaer Polytechnic Institute, Department of Computer Science, USA

Abstract: The promise of stem cell-based therapy is predicated on harnessing the plasticity of stem cell phenotypes to repair or replace damaged tissues. As technologies for detecting, isolating, modifying, tracking, and even inducing stem cells improve, the very definition of what constitutes a stem cell is now an open question. Addressing this fundamental problem has triggered an explosion of activity that spans the entire breadth of biological fields, from molecular biology to population biology. While this has clearly increased the gross amount of information concerning stem cells, its net impact is limited by a lack of integrative multiscale models that are readily accessible to researchers from many disciplines. The field of embryonic stem (ES) cell biology is a good example of the strengths and limitations of the segregative reductionist approach. The goal of this brief review is to highlight some of the most promising recent advances in embryonic stem cell research, with an emphasis on how data gathered from one level can benefit research across multiple scales.

Keywords: Multiscale modeling, embryonic stem cells, concurrent methods, hierarchical methods, systems biology, design optimization, feature selection, cell-graphs, induced pluripotent stem cells, supervised learning, machine learning, stem cell niche, imaging, graph theory, tissue modeling Tissue structure/function, cell-cell communication.

MULTISCALE MODELING AS AN INTEGRATIVE TOOL

The high expectations generated by stem cells and their potential clinical applications have resulted in an alignment of several biological disciplines around a central purpose. While this has been a positive force for streamlining collaboration between clinicians and basic scientists, it has also exposed gaps in

*Address correspondence to B. Yener: Rensselaer Polytechnic Institute, Department of Computer Science, Lally 310, 110 Eighth Street Troy, NY 12180-3590, USA; Tel: 518-276-6907; Fax: 518-276-4033; E-mail: yener@cs.rpi.edu

our ability to synthesize this information into a coherent whole. In part, this reflects the historical subdivision of biomedical sciences into discrete fields that by necessity establish their own jargon and measures of "success." While a great deal of effort is committed to achieving these successes, all too often it fails to impact other fields aiming for the same target, stem cell therapy. In many ways, the current situation in stem cell biology resembles that of the mid-1980s in the cancer field, when bench-to-bedside research was more concept than reality [1].

While there are many complex issues slowing the entry of stem cell technology into the clinic, one of the most significant issues is our inability to understand and predict the behavior of stem cells based on their characteristics at either the molecular, cellular, or tissue level. For stem cells to make their way into clinical applications, it is critical that we develop methods for predicting stem cell behavior. Multiscale modeling has emerged as a powerful means to integrate research across many levels, from molecular structure to potential therapeutic interventions [2, 3]. These multiscale models, which include information from multiple levels of analysis, also permit the generation and testing of hypotheses, creating an iterative systems biology loop [4]. The scales can span from molecular to organ and systems levels. The level of sophistication that multiscale models have been able to achieve now makes them a useful tool to apply towards the study of stem cells.

Most existing models are limited to one biological scale, or level. Modeling techniques at the molecular level aim to capture biochemical activities to achieve a comprehensive quantitative understanding of molecular interactions. However, these simulations are not computationally scalable and are limited to simulating only a few million atoms over a few nano-seconds. Hence, molecular and sub-cellular level models, such as Virtual-cell (http://www.nrcam.uchc.edu/), are not capable of describing even a single cell completely. Likewise, modeling techniques at the organ or systems level capture coarse grained properties by summarizing or averaging the quantification of tissue and cellular level activity but fail to capture fine-grained molecular details. Such models typically are physics-based and typically use algebraic and parametric optimization techniques rather than physical data. To accurately model any biological system, including stem cells, requires integration across all of these biological scales, to produce a multiscale model. Multiscale models provide a computational challenge, since achieving a computationally accurate continuum requires seamless integration of the data, models, and the knowledge from different scales. This integration is achieved computationally through scale linking.

Scale linking is currently performed using hierarchical and concurrent schemes [5]. In *concurrent methods*, at least two distinct models (*e.g.,* atomistic and continuum descriptions) are used simultaneously in the same simulation with special "hand-shake" regions for information passing. These methods are used when the need to account for the finer scale (more expensive model) is limited to a small portion of the domain. In *hierarchical methods* information from one scale is passed to the other sequentially at sampling points typically distributed through the domain of the problem. There is an intuitive compatibility between image-based systems and hierarchical methods, while the concurrent methods are suitable with physics-based approaches.

Mathematically sound single-scale models that are computationally stable and statistically meaningful demand more data than most biology laboratories and clinics can provide. This disparity represents a fundamental bottleneck in statistically rigorous biomedical research. That is, as the data becomes more relevant and important from the modeling point of view, it becomes more expensive and less available. For example, the cost of generating *in vitro* data is at least two orders of magnitude less than generating the same data from clinical samples - hence the proliferation of *in vitro* drug screens. Thus the data bottleneck problem is defined by the difficulty of finding the most relevant data (*i.e.,* human sources, animal models, and 3D tissue/organ explants as supposed to *in vitro*) that can be obtained under various control schemes (possible for *in vitro* and *in vivo* models but not for human sources) at high volume and at low cost. Thus, a multiscale approach is necessary to widen the data bottleneck, integrate multiple sources of data and knowledge, and bridge the computational and biological sciences to develop systems biology loops.

To accelerate discovery, systems biology loops learn more quickly than traditional trial-and-error approaches, intelligently shrinking the experimental search space to identify high-priority experiments. All systems biology approaches focus on defining three characteristics of biological systems: *robustness* (ability to maintain phenotypic stability in response to perturbation), *modularity* (clustering of components into functional "teams"), and, most importantly, *emergent properties* (behaviors unique to the entire system, and not found from the summation of its constituent parts) [6]. At present, all three of these characteristics remain largely unknown for most stem cells.

Effective models often result from close collaboration between wet-bench experimentalists and computation experts [7], who bring the rigor of predictive modeling strategies from a variety of disciplines, ranging from advertising campaigns to manufacturing processes that ensure proper seasoning of snack foods [8]. For many biologists, the key to developing such a multiscale model is to convert a medical goal into a *design optimization* problem. Like homeostasis, design optimization seeks to achieve an "optimal set point" of input values to yield a defined output. For example, if stem cell biologists seek to stimulate an ES cell to adopt a specific differentiated phenotype, the design optimization problem is to select which input variables in the ES environment (*e.g.,* growth factor concentration, extracellular matrix composition and organization, application of mechanical force) should be changed, and by how much, to achieve that phenotype, which is defined by a set of features (commonly referred to as a "gold standard"). These features may include gene/protein expression, cell morphology, cell-cell and cell-matrix adhesion, *etc.*, or more complex metrics of organizational and functional complexity. Optimization refers to varying the ES environment inputs until the values of these features (collectively called a "profile") in the ES cells match the values of the gold standard. Four stages are required: (1) initial data is captured to yield a profile of feature values in the ES cells, (2) the predictive model, built from profiles, is used to suggest new settings in the input values to push the cells toward the gold standard profile, (3) the models are tested directly to determine how effectively the given inputs drive the cells towards the gold standard, and then (4) this new data is used to improve the model, suggest additional "optimized" input changes, and ultimately yield a better product. With each cycle of inputs and outputs, the model moves closer to achieving the desired product. The challenge is to define the "gold standard" for such optimization, since each scale of stem cell biology focuses on its own set of features.

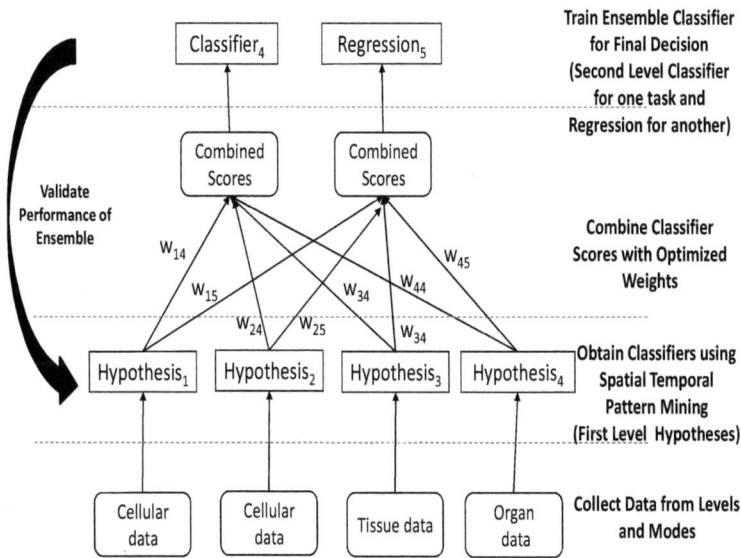

Figure 1: Architecture of a multiscale many-to-many learning system. An ensemble classifier is used to maximize the accuracy of predicting the outcome of the underlying biological processes.

An analytical way of addressing this optimization problem is to extract quantifiable features or metrics at each scale and then determine which one of them best describes biological activity at that particular scale. This can be done using *feature selection* methods [9]. Based on the features, supervised machine learning techniques such as Support Vector Machines (SVD) can be deployed to classify and predict the structure-function relationship at that scale. Once each scale is modeled, hierarchical learning techniques, such as Ensemble Learning [10], can be used to integrate the scales [11] (see Fig. **1**). The optimization problem then becomes finding the optimal weights of influence of each model at each scale to maximize the prediction accuracy, which can be tested against labeled data. The set of features and the distribution of their values they take can be used to define the gold standard in a quantitative way.

Critical information that is required to generate useful multiscale models at the level of tissue or organ function that also incorporate molecular and cellular detail is image-based data. A method to obtain quantitative information on cellular organization is the application of graph theory to images. Graph theory can be used for representation of cellular, tissue, and organ level organization [12-16] to provide a rich set of features at different scales, as shown in Table **1**. In these studies, cell-graphs have been developed from images in which cells are labeled with a nuclear marker and are recognized as "nodes" or "points" in the cell-graphs and "links" or "edges" are established based on pairwise distances. Cell-graphs have been useful for identifying quantitative differences in tissue organization between normal and cancer cells [12-15]. These studies have been extended to bone tissue remodeling [17] and to classification of different cell types in developing salivary glands [18]. Therefore, it is likely that cell-graphs will also provide a useful tool for quantitative analysis of stem cell organization. Explanations of these metrics can be found in [12-19].

To obtain quantitative spatial molecular information requires other quantitative methods, such as image segmentation. With image segmentation methods, molecules can be traced to subcellular regions within specific cells [19, 20]. Although image acquisition is traditionally a low-throughput method, use of tissues arranged into arrays to create tissue arrays, facilitates automated acquisition of images from multiple samples nearly simultaneously. The combination of tissue microarrays with quantitative image segmentation methods has been useful to identify changes in the subcellular localization of the receptor protein Met as a prognostic indicator in stage I and stage II colon cancer [21]. Future application of image analysis methods will provide insight into understanding the complex interactions of stem cells with their environment.

Table 1: Features extracted from images analysis and using cell-graphs at the cell, tissue, and organ level.

Feature	Cell	Tissue	Organ
(Average) Degree	x	x	x
Clustering Coefficient C, D, E	x	x	
Average Eccentricity		x	x
Diameter/radius		x	x
Average Path Length		x	x
Number o f Connected Components		x	x
Percentage of Isolated Points/of End Points		x	x
Number of/Percentage of Central Points		x	x
Number of vertices / Number of Edges		x	x

Spectral Radius / Spectral Gap/ Trace		x	x
Second Largest Eigen Value		x	x
Normalized Laplacian Number of 0s, Number of 1s		x	x
Normalized Laplacian Lower (Upper) Slope/Tace/Energy		x	
Elongation	x	x	x
Area	x	x	x
Orientation	x	x	x
Eccentricity	x	x	x
Perimeter	x	x	x
Circularity	x	x	x
Convexity	x	x	x

STEM CELLS ARE DIFFICULT TO DEFINE

The stem cell concept was first proposed by Alexander Maximov in 1906 to describe a cell that is capable of reconstituting hematopoiesis in bone marrow [22]. His ideas were experimentally confirmed in 1963 [23]. Since that time, the definition of a stem cell has expanded and evolved to include several different perspectives. To the clinician, a stem cell has the ability to replicate and differentiate to regenerate a healthy replacement for damaged tissue; to the cell or molecular biologist, a stem cell undergoes asymmetric cell divisions to generate two daughter cells with distinct gene and protein expression profiles, such that one retains its stem cell characteristics, while the other commits to establishing a mature, fully differentiated phenotype by turning on a differentiation program [24]. For tissue engineers, stem cells are those cells capable of replacing an entire adult tissue *de novo*, either by generating the necessary cells themselves or recruiting them from elsewhere in the body.

While any "true" stem cell may possess each of these traits, plus others, there is currently a lack of consensus. Indeed, the plethora of proposed "defining characters" in stem cells has caused some to question whether any single cell in the human body qualifies [25]. Current classification of stem cells is primarily focused on tissue of origin and/or differentiation potential. At present, the ~20 known human stem cells are grouped into classes based on their differentiation potential: pluripotent stem cells possess the ability to differentiate into any adult tissue type, while multipotent cells possess a more limited differentiation potential. Most stem cells are multipotent, capable of generating a handful of cell types that function together in the same tissue. This "all-or-some" boundary is helpful for discriminating between embryonic stem cells and most others, but offers no mechanistic insight as to how "stemness" is generated or controlled.

Embryonic stem cells are now largely defined by their location in developing embryos (the inner cell mass of the blastocyst) [26], and their ability to regenerate cell types derived from all three germ layers. The recent demonstration that fibroblasts and other adult somatic cells can be reprogrammed to resemble pluripotent embryonic stem cells by inducing expression of four transcription factors [27] has virtually erased the word "terminal" from all descriptions of differentiated cells to blur the line between pluripotency and differentiation.

From a design optimization standpoint, to achieve a successful stem cell therapy approach an absolute definition of a stem cell is not necessary; one merely needs to identify a quantitative goal, such as achieving a specific gene expression pattern or specificity of homing to a defined target tissue. In this light, the abundance of "discriminative features" is an advantage, because it provides a greater coverage of stem cell phenotype. As an example, we will examine the embryonic stem cell field at several levels to identify possible inputs to such a model.

MOLECULAR LEVEL: CAN WE OPTIMIZE A MOLECULAR PROFILE?

Pluripotent embryonic stem cells were first cultured from the inner cell mass of mouse blastocysts in 1981 by two groups working independently [28, 29]. The establishment of the first cultured human embryonic stem cells from blastocysts eventually followed in 1998 [30]. The human cell lines initiated an ethical debate that lead to a ban on use of Federal dollars to fund human stem cell research in August of 2001, a restriction that was eased in March, 2009. The ethical complaints arise from the destruction of the embryo that is required to create an ES stem cell line. This led to the finding that ES cell lines could be created from a single blastomere without killing an embryo [31] and investigation into other stem cell sources, such as in adult tissues. While stem cells are found in adult tissues and presumably in any tissue that undergoes renewal, the potency of such stem cells is at best multipotent. Nevertheless, much work has been done to investigate the use of adult stem cells for therapeutic purposes. The comparison of adult stem cells with embryonic stem cells to understand the molecular differences underlying the functional differences between these types of cells is another question of interest. Unfortunately, due to the technical restrictions on numbers of markers that can be simultaneously investigated, the lack of a standard marker set, and the variable results that have been obtained from such work, it is difficult to compare results from many of these studies.

With the finding in 2006 that mouse fibroblast cells can be "reprogrammed" into a pluripotent state following the introduction of four transcription factors into the cells [27, 28] and the subsequent finding that the same methods were effective for reprogramming human fibroblasts [32], a new interest has arisen in using these induced pluripotent stem (iPS) cells for therapeutic options. These findings have opened up the possibility that it will one day be possible to remove somatic cells from a patient, reprogram them, possibly fix a genetic defect in the cells (if necessary) and then re-implant them into the patient. With this possible new technology, many new questions have been raised, including: What are the differences between the pluripotent state of iPS cells and human ES cells? How can iPS cells be controlled *in vivo* to prevent development of cancer? What is the difference between partially reprogrammed and fully reprogrammed cells, and can these partially reprogrammed cells be used therapeutically instead? Investigation into these issues has led to the realization that more rigorous characterization of stem cells is needed and that standardization of markers is critical prior to the entry of iPS cells into the clinic [33].

While gene expression profiling and proteomic profiling have both been used to attempt to define stem cell markers, there has so far been no absolute consensus reached on the definition of the pluripotent phenotype. To further complicate matters, the signal transduction pathways controlling gene expression and protein turnover are themselves dynamic and linked by complex regulatory mechanisms that are poorly understood. A greater understanding of the pathways controlling pluripotency and the integration of these pathways is greatly needed.

How can computational methods help clarify these issues? In the absence of the ground truth, or knowledge of all the properties of stem cells, supervised learning techniques are not applicable to discover the pathways controlling pluripotency. Thus, unsupervised learning techniques, such as Singular Value Decompositions (SVD), Principal Component Analysis (PCA), Multidimensional Scaling (MDS), and Independent Component Analysis (ICA) would be applicable. Unsupervised learning algorithms can help by providing density estimation and dimensionality reduction. Density estimation aims to learn the parameters of a probabilistic model for prediction, while the aim of dimensionality reduction is to create more compact representations of the original data, while capturing the information necessary for knowledge extraction. However, these unsupervised learning techniques work on 2D, tabular (*i.e.,* matrix) representations of data and are primarily suited for bilinear models. Thus, generalization of techniques such as SVD from 2-way models (matrices) to 3-way (cubes), and to multi-way modeling and analysis [34, 35] is needed for multilinear data (*e.g.,* time series of data matrices). Rank reduction techniques such as Tucker [36] or PARAFAC [37] are well suited for a 3-way model (*e.g.,* gene expression x cellular stimulus x time) [38, 39]. In both multi-linear and bi-linear techniques, the common assumption is that there is a linear structure in the high dimensional data which may not be valid. Thus, techniques such as locally linear embedding (LLE) should be considered for the problem of nonlinear dimensionality reduction [40].

One of the most pressing questions for defining ESCs at the molecular level is how specific environmental stimuli, signaling molecules, and gene regulatory networks cooperate to control growth and differentiation of these cells. Multiscale models show great promise in addressing such questions. For example, deterministic, probabilistic, and statistical learning models are used to extract information about the functional changes in proteomic networks that may define a stem cell phenotype [41]. The stochastic nature of proteomic, genomic, and signaling data suggests that other machine learning methods, such as Bayes Networks, can also be used to model proteomic networks such as protein-protein interactions. Doug Lauffenburger's group at Massachusetts Institute of Technology pioneered the use of systems biology with stem cells [42, 43], and this has now expanded into a robust research field [44-46]. One key component of this strategy is to adopt macro-scale approaches to gather large amounts of molecular data. For example, commercial protein phosphorylation arrays capture the signaling behavior of dozens of protein kinases controlling cell growth and differentiation, and can be combined with proteomic and DNA microarrays to capture a comprehensive picture of cellular phenotype in these studies [47]. These high-throughput techniques are particularly useful for near homogeneous cell populations.

Predictions from multiscale models can be tested experimentally using existing techniques to generate mechanistic relationships between these molecules. This would benefit not only those interested in pinning down the discriminative profile (s) of ESC, but would provide the necessary gold standards for optimization of the culture conditions for expanding and inducing differentiation of these cells. This is how data from one level of stem cell research can positively impact others.

CELLULAR LEVEL: CAN WE DEFINE THE ESC NICHE?

Like the stem cell itself, the notion of a specific environment that controls its growth and preserves its undifferentiated state, or niche, emerged as a concept before its existence was positively demonstrated [48]. The molecular makeup and regulatory mechanisms of most stem cell niches, including that of ESCs, are almost entirely unknown. Extrapolating from what we know about hematopoietic and reproductive stem cell niches in animal models, the ESC niche is likely defined by three features: 1). cell-cell adhesions to both anchor stem cells and signal their self-renewal, 2). physical organization that provides spatial and regulatory cues and 3). soluble signals that regulate maintenance. These three features are maintained by heterologous cell populations, the extracellular matrix, and paracrine signaling, respectively.

How do we find/define these features? Due to the elusive nature of the stem cell niche and its small, locally confined and yet complex environment, standard molecular and biochemical profiling approaches are not suited to identifying its features. A promising approach to finding the niche is *in vivo* fluorescence microscopy; however, this approach is currently limited in the number of components that can be simultaneously probed. Nevertheless, quantitative analysis of fixed fluorescent images was recently used to identify properties of the neural stem cell niche in the adult subventricular zone [49].

Other approaches are needed to identify the properties of the stem cell niche. These could include laser capture microdissection followed by RNA isolation, amplification and molecular profiling within the isolated tissue; however, such approaches are still limited to mRNA expression and to expression within groups of cells. Physical properties of the stem cell niche are much more difficult to identify, yet these properties will be important for muliscale models. Multiscale models have the potential to use input such as composition and 3D orientation of specific extracellular stimuli to both predict expression of marker proteins and genes but also to help predict the composition of the extracellular environment in the cell niche.

This is precisely the approach used by van Leeuwen *et al.,* [3] to establish links between molecular signaling networks, mechanical properties of cellular interactions, and tissue-level organization of the gut epithelium. Virtual microdissections and labeling experiments revealed that the gut epithelial stem cell niche is dynamic and the stem cells contained within it are mobile, even as they retain their stemness. Such an elegant, multiscale approach linking cellular phenotype with spatiotemporal information could likely assist in the definition of many other stem cell niches, including those for ESCs. Most importantly, this strategy takes full advantage of the genetic, molecular, and histological data focused on characterizing the maintenance of an entire tissue.

TISSUE LEVEL: CAN WE ENGINEER A STEM CELL NICHE *EX VIVO*?

For many investigators, the ultimate goal of stem cell research is to harness stem cells' capacity to reconstitute tissues to treat specific injuries or diseases. *In vitro* culture of intact 3D tissues and reconstitution with tissue extracts and engineered scaffolds are attempts to recreate or fabricate a niche. Most often, such strategies operate largely independent of the pursuit for a definitive molecular stem cell profile. Without a definitive target cell to characterize, those in search of its niche are handicapped from the outset. In some cases, simply injecting ESC into decellularized tissue explants (*e.g.,* [50]) yields promising results. Despite years of work in design of artificial scaffolds, use of deceullarized scaffolds has been most successful *in vivo* [51, 52], largely due to our lack of knowledge of how to engineer an *in vivo* environment. But in most cases, at least partial reconstitution of the proper cellular niche will be required for long-term success of stem cell therapies to facilitate tissue maintenance over time.

Since the late 1980s engineering design principles have been applied to living systems to create replacement tissues *de novo*. In its most basic sense, an engineered tissue construct (ETC) is a three-dimensional assembly of one or more cell types suspended in an extracellular scaffold material and fed by soluble molecules, including growth factors, hormones, and nutrients. Once assembled, the ETCs are intended to be implanted as replacements for damaged or diseased tissues. The results thus far have been promising [53], and some enjoy widespread use in the clinic [54]. While stem cell-based ETCs hold great promise due to their potential for self-renewal, creating a suitable niche *ex vivo* is a major hurdle to widespread stem cell therapy [55].

The pace of *ex vivo* stem cell niche development is hampered by at least two design problems. First is sorting through the enormous number of possible "ingredients" (cells, scaffolds, biochemicals, stimuli). As an example, some groups have identified as many as nine classes of functional parameters for designing musculoskeletal tissues: differential fiber length, *in vivo* force and displacement, variations in relative attachment site locations, loading from adjacent structures, fiber interactions, types of insertion, regional variations in material properties, nonparallel fiber orientations, and complex loading within the structure [56]. Tissue engineers develop "educated guesses" based on prior experience to guide them. Yet this task is too daunting for even the best-informed tissue engineers to solve by sheer brute-force trial and error methods. Second, most educated guesses in tissue engineering are based on data from complex measures of tissue performance. These tests are often difficult, expensive, and time consuming. If one wants to assemble a blood vessel, for example, the only reliable *ex vivo* measures of "functionality" are end-point measures (burst pressure, tensile strength, strain modulus, *etc.*) [57]. Increasing the amount of reliable data through improvements in *in vivo* imaging and computational methods would be a great help.

Multiscale models, such as that developed by van Leeuwen *et al.*, hold great promise for improving our understanding of stem cells and making stem cell therapy a reality. Integrating the wealth of molecular data with performance data and other measures (*e.g.*, multispectral imaging, electrophysiology, *etc.*), can reveal correlations that brute force analysis simply cannot detect. Once these constructs enter clinical trials, clinical outcomes add yet another level of input. The ultimate test of our careful deliberation will take place in stem cell-based clinical trials; the recent report of poorly characterized "fetal neural stem cells" causing a brain tumor after being transplanted in a 13 year old ataxia telangiectasia patient [58] underscores the danger of moving forward without due diligence.

CONFLICT OF INTEREST

None declared.

ACKNOWLEDGEMENTS

The authors would like to acknowledge NIH grants 1RO1 AR053231 (to G.P.), R01 EB008016 (to B.Y. and G.P.), NIH 1RO1 DE192444 (to M.L. and B.Y.), and 1RC1DE020402 (to M.L.) for partial support of this work.

REFERENCES

[1]   Siminovitch L. Advances in cancer research: bench to bedside. J Thorac Cardiovasc Surg 1990; 100:874-878.
[2]   Engler AJ, Humbert PO, Wehrle-Haller B, Weaver VM. Multiscale modeling of form and function. Science 2009; 324:208-212.
[3]   van Leeuwen I, Mirams GR, Walter A, *et al.* An integrative computational model for intestinal tissue renewal. Cell Prolif 2009; 42:617–636.
[4]   Schadt EE, Zhang B, Zhu J. Advances in systems biology are enhancing our understanding of disease and moving us closer to novel disease treatments. Genetica 2009; 136:259-269.
[5]   Yip S, ed., Handbook of Materials Modeling. Dordrecht: Springer 2005.
[6]   Aderem, A. Systems biology: its practice and challenges. Cell 2005; 121:511-513.
[7]   Liu ET. Systems biology, integrative biology, predictive biology. Cell 2005; 121:505-506.
[8]   Yu H, MacGregor JF. Multivariate image analysis and regression for prediction of coating content and distribution in the production of snack foods. Chemometrics and Intelligent Laboratory Systems 2003; 67:125-144.
[9]   Liu H, Motoda, H. Feature Selection for Knowledge Discovery and Data Mining. Norwell MA: Kluwer Academic Publishers;1998.

[10]   Alpaydin E. Introduction to Machine Learning. Cambridge, MA: The MIT Press; 2004.

[11]   Kaynak C, Alpaydin E. Multistage cascading of multiple classifiers: One man's noise is another man's data. 17th International Conference on Machine Learning, ed. P. Langley, 455-462. San Francisco: Morgan Kaufmann;2000.

[12]   Gunduz C, Yener B, Gultekin SH. The cell graphs of cancer. Bioinformatics 2004; 20:i145-i151.

[13]   Demir C, Gultekin SH, Yener B. Augmented cell-graphs for automated cancer diagnosis, Bioinformatics. 2005; 21 (Suppl 2): ii7-ii12.

[14]   Demir C, Gultekin SH, Yener B. Learning the topological properties of brain tumors, IEEE/ACM Transactions on Computational Biology and Bioinformatics 2005; 2 (3):262-270.

[15]   Bilgin CC, Demir C, Nagi C, Yener B. Cell-graph mining for breast tissue modeling and classification. Engineering in Medicine and Biology Society, 29th Annual International Conference of the IEEE. 2007; 5311-5314.

[16]   Lund AW, Bilgin C, Al Hasan M, McKeen L, Stegemann JP, Yener B, Zaki M, Plopper GE. Quantification of spatial parameters in 3D cellular constructs using graph theory. Journal of Biomedicine and Biotechnology 2009; doi:10.1155/2009/928286.

[17]   Bilgin CC, Bullough P, Plopper GE, Yener B. ECM-Aware cell-graph mining for bone tissue modeling and classification. Journal of Data Mining and Knowledge Discovery 2009; 20(3): 416-438.

[18]   Bilgin CC, Ray S, Baydil B, Daley WP, Larsen M, Yener B. Multiscale feature analysis of salivary gland branching morphogenesis. PLoS One. 2012; 7(3): e32906. doi: 10.1371/journal.pone.0032906.

[19]   Can, A, *et al.,* Techniques for cellular and tissue-based image quantification of protein biomarkers, in Microscopic Image Analysis for Lifescience Applications, J. Rittscher, R. Machiraju, and S.T.C. Wong, Editors. New York: Artech House; 2008.

[20]   Gurcan M, Boucheron L, Can A, Madabhushi A, Rajpoot N, Yener B. Histopathological image analysis: A review, IEEE reviews in Biomedical Engineering Vol 2, 2009.

[21]   Ginty, F., Adak, S., Can, A., Gerdes, G., Larsen, M., Cline, H., Filkins, R., Pang, Z, Li, Q., and Montalto, MC. The relative distribution of membranous and cytoplasmic met is a prognostic indicator in stage I and II colon cancer. Clin Cancer Res 2008; 14 (12):3814-3822.

[22]   Maximow AA. Über experimentelle Erzeugung von Knochenmarks-Gewebe. Anatomischer Anzeiger 1906; 28:24-38.

[23]   Becker AJ, Culloch EA, ll JE Cytological demonstration of the clonal nature of spleen colonies derived from transplanted mouse marrow cells. Nature 1963; 197:452-454.

[24]   Orford KW, Scadden DT. Deconstructing stem cell self-renewal: genetic insights into cell-cycle regulation. Nat Rev Genet 2008; 9:115-128.

[25]   Parker GC, Anastassova-Kristeva M, Eisenberg LM, Rao MS, Williams, MA, Sanberg PR, English D. Stem cells: shibboleths of development, part II: Toward a functional definition. Stem Cells Dev 2005; 14:463-469.

[26]   Gavrilov S, Papaioannou VE, Landry DW. Alternative strategies for the derivation of human embryonic stem cell lines and the role of dead embryos. Curr Stem Cell Res Ther 2009; 4:81-86.

[27]   Takahashi K, Yamanaka S. Induction of pluripotent stem cells from mouse embryonic and adult fibroblast cultures by defined factors. Cell 2006;126 (4):663-76.

[28] Evans MJ, Kaufman MH. Establishment in culture of pluripotential cells from mouse embryos. Nature 1981;292 (5819):154-6.

[29] Martin GR. Isolation of a pluripotent cell line from early mouse embryos cultured in medium conditioned by teratocarcinoma stem cells. Proceedings of the National Academy of Sciences of the United States of America 1981;78 (12):7634-8.

[30] Thomson JA, Itskovitz-Eldor J, Shapiro SS, *et al.* Embryonic stem cell lines derived from human blastocysts. Science 1998;282 (5391):1145-7.

[31] Klimanskaya I, Chung Y, Becker S, Lu SJ, Lanza R. Derivation of human embryonic stem cells from single blastomeres. Nature Protocols 2007; 2 (8):1963-72.

[32] Takahashi K, Tanabe K, Ohnuki M, *et al.* Induction of pluripotent stem cells from adult human fibroblasts by defined factors. Cell 2007;131 (5):861-72.

[33] Chan EM, Ratanasirintrawoot S, Park IH, *et al.* Live cell imaging distinguishes bona fide human iPS cells from partially reprogrammed cells. Nature Biotechnology 2009;27 (11):1033-7.

[34] Smilde AK, Bro R, Geladi P. Multi-way Analysis: Applications in the Chemical Sciences. New York: John Wiley & Sons; 2004.

[35] Acar E, Yener B. Unsupervised Multiway Data Analysis: A Literature Survey. IEEE Transactions on Knowledge and Data Engineering 2009; 21 (1):6-20.

[36] Tucker LR. Some mathematical notes on three-mode factor analysis. Psychometrika, 1966 31:279-311.

[37] Harshman RA. Foundations of the PARAFAC procedure: Models and conditions for an explanatory multi-modal factor analysis. UCLA Working Papers in Phonetics 1970; 16:1-84.

[38] Bennett KP, Bergeron C, Acar E, Klees RF, Vandenberg SL, Yener B, Plopper GE. Proteomics reveals multiple routes to the osteogenic phenotype in mesenchymal stem cells. BMC Genomics 2007; 8:380.

[39] Yener B, Acar E, Aguis P, Bennett KP, Vandenberg S, Plopper GE. Multiway Modeling and Analysis in Stem Cell Systems Biology. BMC-Systems Biology 2008; 2:63.

[40] Roweis ST, Saul LK. Nonlinear dimensionality reduction by locally linear embedding. Science 2000; 290:2323-2326.

[41] Janes KA, Lauffenburger DA. A biological approach to computational models of proteomic networks. Curr Opin Chem Biol 2006; 10:73-80.

[42] Prudhomme W, Daley GQ, Zandstra P, Lauffenburger DA. Multivariate proteomic analysis of murine embryonic stem cell self-renewal *versus* differentiation signaling. Proc Natl Acad Sci USA 2004; 101:2900-2905.

[43] Woolf PJ, Prudhomme W, Daheron L, Daley GQ, Lauffenburger DA. Bayesian analysis of signaling networks governing embryonic stem cell fate decisions. Bioinformatics 2005; 21:741-753.

[44] Foster DV, Foster JG, Huang S, Kauffman SA. A model of sequential branching in hierarchical cell fate determination. J Theor Biol 2009;260 (4):589-97.

[45] Halley JD, Burden FR, Winkler DA. Stem cell decision making and critical-like exploratory networks. Stem Cell Res 2009; 2:165-177.

[46] Murali T, Rivera CG. Network legos: building blocks of cellular wiring diagrams. J Comput Biol 2008; 15:829-844.

[47] Albeck JG, MacBeath G, White FM, Sorger PK, Lauffenburger DA, Gaudet S. Collecting and organizing systematic sets of protein data. Nat Rev Mol Cell Biol 2006; 7:803-812.

[48]   Schofield R. The relationship between the spleen colony-forming cell and the haemopoietic stem cell. Blood Cells; 1978 4:7-25.

[49]   Shen Q., Wang Y., Kokovay E., Lin G., Chuang S-M, Goderie SK, Roysam B., and Temple S. Adult SVZ stem cells lie in a vascular niche: A quantitative analysis of niche cell-cell interactions Cell Stem Cell 2008; 3 (3):289-300.

[50]   Behfar A, Hodgson DM, Zingman LV, Perez-Terzic C, Yamada S, Kane GC, Alekseev AE, Puceat M, Terzic A. Administration of allogenic stem cells dosed to secure cardiogenesis and sustained infarct repair. Ann N Y Acad Sci 2005; 1049:189-198.

[51]   Isch JA, Engum SA, Ruble CA, Davis MM, Grosfeld JL. Patch esophagoplasty using AlloDerm as a tissue scaffold. Journal of Pediatric Surgery 2001; 36 (2):266-8.

[52]   Zhao Y, Zhang S, Zhou J, *et al.* The development of a tissue-engineered artery using decellularized scaffold and autologous ovine mesenchymal stem cells. Biomaterials. 2010; (2):296-307.

[53]   Malchesky PS. Artificial Organs 2008: a year in review. Artif Organs 2009; 33:273-295.

[54]   Shieh SJ, Vacanti JP. State-of-the-art tissue engineering: from tissue engineering to organ building. Surgery 2005; 137:1-7.

[55]   Toyoda M, Takahashi H, Umezawa A. Ways for a mesenchymal stem cell to live on its own: maintaining an undifferentiated state *ex vivo*. Int J Hematol 2007; 86:1-4.

[56]   Butler DL, Shearn JT, Juncosa N, Dressler MR, Hunter SA. Functional tissue engineering parameters toward designing repair and replacement strategies. Clin Orthop Relat Res 2004; S190-S199.

[57]   Nerem RM. Tissue engineering of the vascular system. Vox Sang 2004; 87 Suppl 2:158-160.

[58]   Amariglio N, Hirshberg A, Scheithauer BW, *et al.* Donor-derived brain tumor following neural stem cell transplantation in an ataxia telangiectasia patient. PLoS Med 2009; 6:e1000029.

CHAPTER 5

Computational Analysis of DNA-Methylation and Application to Human Embryonic Stem Cells

Lukas Chavez[*]

Max-Planck-Institute for Molecular Genetics, Berlin, Germany

Abstract: Methylation of cytosines is a reversible and dynamic epigenetic DNA modification emerging during differentiation of human embryonic stem cells (hESCs) and throughout mammalian development. Crucial advancements in sequencing technologies have enabled the analysis of DNA methylation on a full genome level. Several studies recently examined the methylomes of hESCs, and investigated genetic and epigenetic dependencies during early differentiation. Methylated DNA immunoprecipitation (MeDIP) followed by high-throughput sequencing (MeDIP-seq) has become a cost-efficient experimental approach for genome wide epigenetic studies. However, it has been shown that MeDIP-seq data has to be corrected for a DNA sequence composition dependent bias in order to produce valid methylation profiles. Therefore, the development and implementation of time-efficient computational methods able to process large amounts of sequencing data with respect to its inherent complexity, is crucial for reducing the imbalance of sequencing data generation and analysis. This chapter introduces to different experimental techniques available for full genome methylation analysis. Subsequently, time efficient algorithms for processing MeDIP-seq data as well as different concepts for normalization are presented. Finally, recent findings of genetic and epigenetic dependencies in hESCs are summarized.

Keywords: DNA methylation, CpG islands, transcription factor binding sites, immunoprecipitation, epigenetics, differentiation, MeDIP, MeDIP-seq, next generation sequencing, linear model, MEDIPS, regulation, transcriptional regulation, pluripotency, normalization.

1. INTRODUCTION

1.1. DNA Methylation

DNA methylation describes the reversible attachment of a methyl group (CH_3) at the 5' position of the pyrimidine derivate cytosine. DNA methylation does not change the DNA sequence itself, and is therefore considered as an epigenetic

*Address correspondence to Lukas Chavez: Max-Planck-Institute for Molecular Genetics, Department Lehrach (Vertebrate Genomics), Bioinformatics Group Ihnestraße 63-73, 14195 Berlin, Germany; Tel: 0049/30-8413-1743; Fax: 0049/30-8413-1769; E-mail: chavez@molgen.mpg.de

Ming Zhan (Ed)

modification. The process of DNA methylation is mediated by methyltransferase enzymes (see Fig. (**1**)).

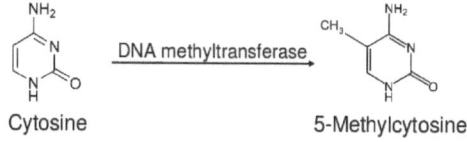

Cytosine   5-Methylcytosine

Figure 1: Cytosine methylation. DNA methyltransferases mediate the reversible attachment of a methyl group (CH3) at the 5' position of cytosines.

Methyltransferase enzymes are commonly classified according to their enzymatic activities. On the one hand, maintenance of methylation activity is necessary to preserve DNA methylation after every cellular DNA replication cycle. On the other hand, *de novo* methylation sets up DNA methylation patterns in early embryonic development [2]. DNA methylation is involved in transcriptional regulation during embryonic differentiation [3] and reprogramming of somatic cells into induced pluripotent stem cells [4, 5]. Aberrant methylation can be associated with severe effects, for example the induction of cancer [6, 7]. Furthermore, distinct genome wide methylation patterns distinguish different cell-types [8, 9]. In mammals, DNA methylation primarily occurs at CpG sites but a recent study has shown that non-CpG methylation accounts for approximately 25% of all methylated cytosines in human embryonic stem cells [10]. It is supposed that methylation affects gene expression by either interfering with binding of transcription factors or modifying chromatin structure to a repressive state [2]. Meissner *et al.,* [3] have shown that methylated CpGs are dynamic epigenetic marks that undergo extensive changes during cellular differentiation. For example, it has been shown that hypermethylation of high-CpG-density promoters (HCPs) leads to irreversible gene silencing [3]. According to these findings, Rakyan *et al.,* [9] observed a negative correlation between DNA methylation and gene expression at high-, but also at medium-, and contrary to previous notions, at even some low-CpG density promoters. Moreover, Rakyan *et al.,* observed that gene-body methylation positively correlates with gene expression [9].

1.2. Methods and Limitations for Detecting DNA Methylation

In principle, there are two major high-throughput methods for detecting DNA methylation. On the one hand, bisulfite based methods can be applied in order to produce DNA methylation information at base resolution. On the other hand, immunoprecipitation based methods are more cost-effective but received methylation levels are of lower resolution.

1.2.1. Bisulfite Sequencing

Treatment of denatured genomic DNA with sodium bisulphite chemically deaminates unmethylated cytosine residues [11, 12]. Subsequently, unmethylated cytosines are converted to uracils. This chemical treatment turns an epigenetic difference into a genetic difference enabling new base specifc DNA methylation detection techniques like bisulfite sequencing or whole genome shotgun bisulfite sequencing (WGSBS). Although whole genome single-base resolution maps have been generated [10, 13] such techniques cannot yet be cost-effective applied to screen large sets of sequences or samples. As an example, the first full genome methylome of hESCs on base resolution was recently reported at a cost of about 1.2 billion sequence reads [10] In order to restrict the full amount of DNA that has to be analyzed, restriction endonucleases can be utilized. The most widely used methylation-sensitive restriction enzymes for DNA methylation studies are HpaII and SmaI [14]. Reduced representation bisulfite sequencing (RRBS) was introduced to reduce sequence redundancy by selecting only some regions of the genome for sequencing by size-fractionation of DNA fragments after BglII digestion [15] or after MspI digestion [3]. These choices of restriction enzymes enrich for CpG-containing segments of the genome but do not target specific regions of interest in the genome [14].

1.2.2. Immunoprecipitation Followed by Sequencing

Methylated DNA immunoprecipitation (MeDIP) uses an antibody specific for methylated cytosines in order to immunocapture methylated genomic fragments [16] (for a detailed description see section (**2.2.**)). Immunoprecipitated methylated DNA fragments can be detected either by tiling arrays (MeDIP-chip) or by next-generation sequencing (MeDIP-seq). Methylation profiles obtained by the MeDIP approach are not base specific but reflect methylation levels on a resolution restricted by the size of the sonicated DNA-fragments after amplification and size selection. However, in contrast to WGSBS or RRBS, the MeDIP approach can be applied in order to obtain cost-effective and full-genome methylation levels without the limitations caused by methylation-sensitive restriction enzymes.

Nevertheless, it has been shown that MeDIP derived data needs to be corrected for local CpG densities in order to estimate valid methylation levels [1, 17, 18]. This effect is caused by varying efficiency of antibody binding and immunoprecipitation dependent on the local density of methylated CpG sites. Especially the analysis of CpG-poor regions has been assumed to be difficult [16, 17]. While there is applicable software available for analyzing MeDIP-Chip data [17, 18], normalization of MeDIP-seq data is in principle solved (BATMAN, [17]) but remained disproportional time-consuming. In fact, processing of MeDIP-seq data from only one full chromosome (*i.e.,* the human chromosome 1) takes approximately three days on a modern-day server when the BATMAN software [17] is applied. Therefore, the major bottleneck of MeDIP-seq based methylation analysis is the time efficient processing of sequencing data with respect to its inherent complexity. Recently, we proposed an alternative method for processing and normalization of MeDIP-seq data [1]. In principle, our method is a simplification of the method developed by Down *et al.,* [17], but we have shown that results are comparable and can be generated in several hours instead of days.

2. EXPERIMENTAL TECHNIQUES

2.1. Second Generation Sequencing

Sequencing of the full human genome was achieved by a technique developed by Frederick Sanger [19]. The high demand for low-cost sequencing has driven the development of high-throughput sequencing technologies that parallelize sequencing processes, producing thousands or millions of sequences at once [20, 21]. High-throughput sequencing technologies are intended to lower the cost of DNA sequencing beyond what is possible with standard dye-terminator methods [22]. A novel high-throughput sequencing method uses bridge PCR for *in vitro* clonal amplification, where fragments are amplified upon primers attached to a solid surface. This technology is used by the Illumina Genome Analyzer (www.illumina.com).

Alternative sample preparation methods allow the sequencing systems to be used for a range of applications including gene expression, ChIP, and MeDIP. After having performed the experiment of interest on the targeted biological material, libraries have to be generated by adapter ligation to the DNA fragments before spread on the flow cells (step 1 in Fig. (**2**)). This step is necessary for the subsequent cluster amplification, as Illumina's sequencing technology relies on

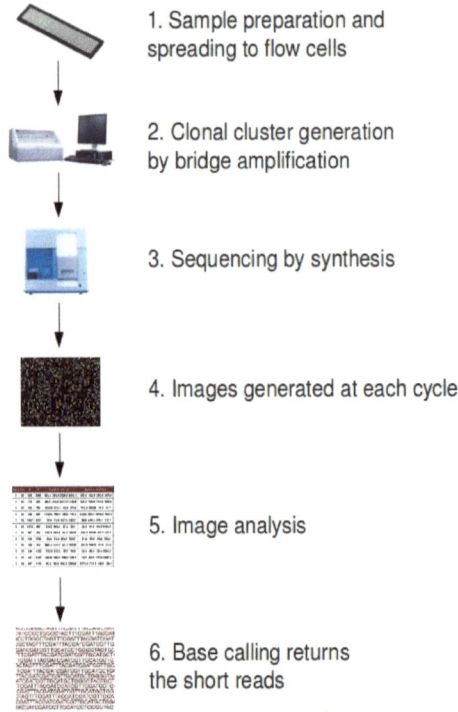

1. Sample preparation and spreading to flow cells

2. Clonal cluster generation by bridge amplification

3. Sequencing by synthesis

4. Images generated at each cycle

5. Image analysis

6. Base calling returns the short reads

Figure 2: Process of Illumina sequencing. The workflow is described in the adjacent text. (Individual images were taken from http://www.illumina.com.)

the attachment of randomly fragmented genomic DNA to a planar surface. Attached DNA fragments are extended and bridge amplified to create high density sequencing flow cells with hundreds of millions of clusters, each containing ~1, 000 copies of the same template (step 2). These templates are sequenced using a four-color DNA sequencing-by-synthesis technology. The sequencing-by-synthesis approach typically runs in 36 cycles (step 3). After each cycle, an image is generated by fluorescence detection (step 4). The first step in the primary analysis is interpreting the image data in order to identify distinct clusters and to create digital intensity files describing the signal intensities of each cluster in each cycle (step 5). Signal intensity profiles for each cluster are used to call bases. Determining the quality of each base call is crucial for downstream analysis and confidence scores for each call are calculated (step 6). Finally, millions of short reads are received, typically of length 36bp. The obtained sequence reads are aligned against the according reference genome and finally application specific data analysis tools are applied. Image analysis, base calling, and efficient quality score dependent alignments are currently intensively explored topics in the field of computational biology.

2.2. Methylated DNA Immunoprecipitation (MeDIP)

Methylated DNA immunoprecipitation (MeDIP) uses an antibody specific for methylated cytosines in order to immunocapture methylated genomic fragments [16]. First, genomic DNA is extracted from cells. Then purified DNA is subjected to sonication in order to shear it into random fragments (see Fig. **3**).

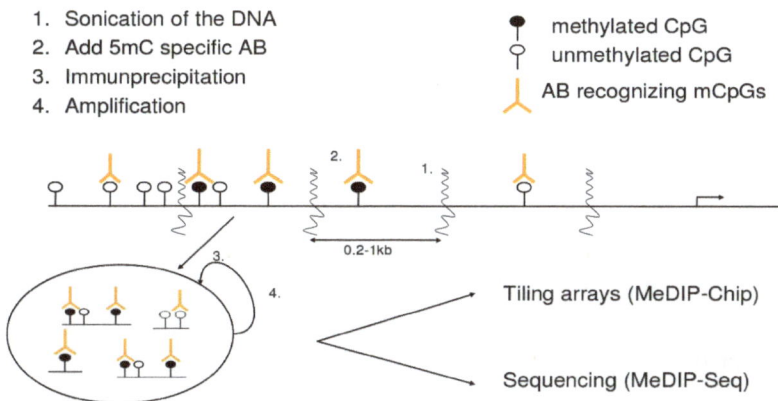

Figure 3: MeDIP workflow overview. In principle, the workflow follows classical ChIP approaches but here, an antibody specific for methylated cytosines (5mC) is utilized. The figure illustrates the varying efficiency of antibody binding and immunoprecipitation dependent on the local density of methylated CpG sites. Especially the analysis of mCpG-poor genomic regions has to be considered critical   (1). MeDIP is followed either by array-hybridization or by second generation sequencing (see also section (**2.1.**)).

The short length of these fragments is important for obtaining adequate resolution, improving the efficiency of downstream steps in immunoprecipitation, and reducing fragment-length effects or biases [23, 24]. To further improve binding affinity of antibodies, the DNA fragments are denatured to produce single-stranded DNA. Following denaturation, the DNA is incubated with monoclonal antibodies which bind to methylated cytosines. Subsequently, classical immunoprecipitation technique is applied. DNA is purified from antibodies by enzymatic digestion and afterwards prepared for DNA detection [25-27]. Analogous to ChIP, immunoprecipitated methylated DNA fragments can be detected either by tiling arrays (MeDIP-Chip) or by next-generation sequencing (MeDIP-seq, see section (**2.1.**)). As already mentioned in section (**1.2.2.**), it has been shown that MeDIP derived data needs to be corrected for local CpG densities in order to estimate valid methylation levels [17, 18].

3. MODELLING OF MEDIP-SEQ DATA

Methylated DNA immunoprecipitation (see section (**2.2.**)) followed by sequencing (see section (**2.1.**)) results in millions of experiment specific short DNA sequences. In order to identify their genomic origin, they are aligned to the according reference genome by applying available alignment implementations like *e.g.,* MAQ [28] or Bowtie [29]. Furthermore, standard post-processing of the alignment results is to filter out mapped reads of low quality and to exclude artificial short read pile-ups. For all remaining short reads, their genomic coordinates together with their associated strand information (plus or minus strand) are extracted and serve as the basic information obtained from a MeDIP-seq experiment.

3.1. Genome Vector

In order to calculate the genome-wide short read coverage, a targeted data resolution has to be determined. In principle, a short read coverage can be calculated for each base position. Because the resolution of MeDIP-seq data is restricted by the size of the sonicated DNA fragments after amplification and size selection (typically between 0.2-1kb), a bin size of 50bp is considered as a reasonable compromise on data resolution and computational costs. Moreover, short reads generated by modern-day sequencers do not represent the full DNA fragments but are of shorter length (*e.g.,* 36bp). Therefore, the data is smoothed by extending each read to a length according to the estimated average length of sequenced DNA fragments, either along the plus or along the minus strand, as specified by the short read dependent strand information. Each chromosome is then divided into bins of size 50bp and the short read coverage is calculated on this resolution. In the following, the bin representation of the genome is called the genome vector (see Fig. (**4**)).

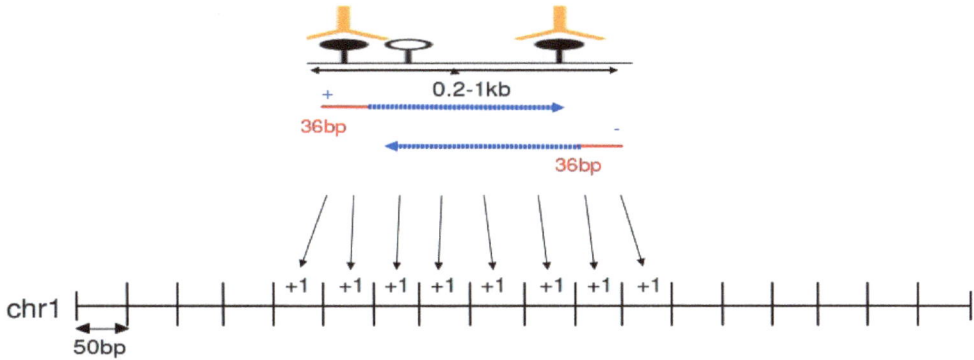

Figure 4: The genome vector. MeDIP-seq derived short reads (red lines) are mapped against the according reference genome. Based on their obtained genomic coordinates, they are extended to a length of *e.g.,* 400bp along the plus or the minus strand according to the associated strand information (blue dashed lines). Genome wide short read coverage is calculated by first defining a targeted resolution (here 50bp bins) and by counting the number of overlapping extended reads at each genomic bin position. The genome vector is the computational representation of the short read coverage at a 50bp bin resolution (single chromosome vectors are concatenated to one genome vector).

Second generation sequencing approaches generate millions of short reads per experiment. Long-term data storage and efficient data processing are challenging tasks, even for modern-day servers. Naive programming approaches cannot be applied in appropriate time. For example, the task of identifying overlapping extended short reads at genome wide 50bp bins needs to be implemented in a sophisticated way.

Algorithm (**1**) shows example R (www.R-project.org) source code for calculating the short read coverage. In fact, this function is called from a wrapper function that processes the chromosomes of the reference genome iteratively. First, the wrapper function declares and initializes the genome vector with respect to the number and lengths of chromosomes of the reference genome and with respect to the targeted coverage resolution (*e.g.,* 50bp). In each iteration step, the wrapper function provides separated vectors for the start, stop, and strand information of the reads (all of the same length *n*, where *n* is the number of short reads available for the current chromosome), as well as a vector containing the chromosomal positions of the predefined bins within the current chromosome. The latter vector is of length $m = \dfrac{length(chromosome)}{binSize}$, defined by simply dividing the length of the current chromosome by the targeted resolution (*e.g., binSize*=50bp). First of all, the short reads are extended (smoothing) to a length as specified by the *extend* parameter. Moreover, for each of the provided vectors, additional identification (*id*) vectors are generated. For the vector containing the start positions of the short reads, an *id* vector of length n is defined (here: *reads_start_id*) and each entry is assigned to the value 1. For the vector containing the stop positions of the short reads, an *id* vector of length n is defined (here: *reads_stop_id*) and each entry is assigned to the value -1. For the vector containing the chromosomal bin positions, an *id* vector of length m is defined (here: *positions_id*) and each entry is assigned to the value 0. In addition, two vectors *ct.vec_pos* and *ct.vec_id*, both of length $2n+m$, are declared. The start and stop position of the short reads as well as the bin positions are concatenated and assigned to the *ct.vec_pos* vector. Additionally, the previously created *id* vectors are concatenated in the same order as the position vectors and are assigned to the *ct.vec_id* vector. This combined *ct.vec_id* vector is sorted with respect to the order of the sorted position vector *ct.vec_pos*. From the algorithmic point of view, this sorting is the most time consuming step that can be solved in an average runtime of $O((2n+m)\log(2n+m))$ (*e.g.,* using quicksort). Next, R's *cumsum ()* function is applied to the ordered *ct.vec_id* vector. The cumulative sum is a sequence of partial sums of a given sequence. The *cumsum ()* function returns a vector of length $2n+m$, where each entry j results by the cumulative sum along the ordered *ct.vec_id* vector. It is obvious that this calculation can be achieved in $O(2n+m)$ calculation steps. The resulting *count* vector is considered as a counter that starts at zero. The *cumsum ()* function increases the counter by 1, whenever a short read starts and decreases the counter by 1, whenever a short read stops. Because the chromosomal bin positions are associated to zeros, the *cumsum ()* function does not change the counter at that positions. The chromosomal bin positions were sorted in between all start and stop positions of the short reads, and therefore, the current value of the counter at the chromosomal bin positions reflects the number of 'open' or overlapping short reads. From the resulting *count* vector, the number of overlapping short reads at each chromosomal bin position can be directly sorted out by selecting all entries whose indices are associated to zeros in the ordered *ct.vec_id* vector.

Algorithm 1: *Calculating short read coverage in* $\mathrm{O}((2n+m)\log(2n+m))$.

```
distributeReads<-function (reads_start = NULL, reads_stop = NULL, reads_strand = NULL,
positions = NULL, extend = 0){

        if (extend != 0){
        reads_start [reads_strand == "-"]=reads_stop [reads_strand == "-"] - extend
        reads_stop [reads_strand == "+"]=reads_start [reads_strand == "+"] + extend
        }

        reads_start_id = vector (length = length (reads_start), mode = "numeric")
        reads_start_id [] = 1

        reads_stop_id = vector (length = length (reads_stop), mode = "numeric")
        reads_stop_id [] = -1

        positions_id = vector (length = length (positions), mode = "numeric")
        positions_id [] = 0

        ct.vec_pos = vector (length = length (reads_start) + length (reads_stop) + length (positions),
mode = "numeric")
        ct.vec_id = vector (length = length (reads_start) + length (reads_stop) + length (positions),
mode = "numeric")

        ct.vec_pos [] = append (append (reads_start, reads_stop), positions)
        ct.vec_id [] = append (append (reads_start_id, reads_stop_id), positions_id)

        ct.vec_id = ct.vec_id [order (ct.vec_pos)]
        count = cumsum (ct.vec_id)

        return (count [ct.vec_id == 0])
```

By this implementation, the task of identifying all overlapping short reads at arbitrary chromosomal bin positions is limited by $\mathrm{O}((2n+m)\log(2n+m))$ in time. Naive approaches easily end up in $2nm$ calculation steps when comparing all start and end positions of the short reads to the chromosomal bin positions. When considering MeDIP-seq data, n has to be at least $2 \cdot 10^7$ for a minimal covered human methylome. As the human genome consists of approximately $3 \cdot 10^8$ base pairs, there will be $m = 6 \cdot 10^7$ genomic bins when defining a bin size of 50bp. Therefore, the presented implementation is limited by $\mathrm{O}((2n+m)\log(2n+m)) = 1{,}842{,}068{,}074$ calculation steps, whereas the naive approach will need $2nm = 2{,}400{,}000{,}000{,}000{,}000$ calculation steps. This example emphasizes the need for time efficient implementations when high-throughput sequencing data is modeled.

3.2. MeDIP-seq Data Normalization

The idea of a MeDIP experiment is to identify cytosine methylation profiles of a sample of interest by immunocapturing methylated cytosines using a mCpG specific antibody [16]. However, it has been shown [17, 18] that MeDIP signals scale with local densities of CpGs and are not necessarily influenced by methylated cytosines, only. Currently, there are two software solutions available for analyzing MeDIP-seq data [1, 17]. Both methods are based on the concept of coupling factors introduced by Down *et al.* [17].

3.2.1. Coupling Factors

MeDIP-seq data normalization approaches [1, 17, 18] incorporate local CpG densities into the MeDIP derived signals. In order to integrate the information on CpG densities into the data, it is necessary to identify the genomic positions of all CpGs. Following the valuable concept of coupling factors presented by Down *et al.,* [17], a *coupling vector* has to be calculated based on the genomic positions of all CpGs. The coupling vector is of the same size as the predefined genome vector but contains local CpG denisties (also called coupling factors) for each genomic bin, instead. For each predefined genomic bin at position b, the density of surrounding CpGs has to be calculated. For this, first a maximal distance (d) has to be defined. Only CpGs within the range of $[b-d, b+d]$ will contribute to the final local coupling factor at b. The optimized value for d will reflect the estimated size of the sonicated DNA fragments after amplification and size selection. This is because MeDIP-seq derived signals at position b are influenced by sequenced DNA fragments that overlap with position b. Immunoprecipitation of these DNA fragments can be caused by a methylated and antibody bound CpG located at any position of the DNA-fragment. The maximal distance of a CpG contributing to the signal at b is therefore the estimated average length of the sonicated DNA fragments (d). Let c be the chromosomal position of a CpG and as b is the chromosomal position of a genomic bin, $dist = |b-c|$ is the distance between the genomic bin and the CpG. A CpG will contribute to the coupling factor of a genomic bin at position b, if $dist \leq d$. The simplest way for calculating local CpG densities is to count the number of CpGs within the maximal distance d around a genomic bin at position b. Another approach is to weight each CpG by its distance to the current genomic bin. CpGs farther away from the current genomic bin will receive smaller weights, whereas CpGs close to the genomic bin will receive higher weights. Fig. (**5**) illustrates a genome vector generated by defining a bin size of 50bp. In addition, CpGs are given in a schematic way. The figure illustrates that immuoprecipitated DNA fragments of an estimated average length greater than the pre-defined bin size can contribute to the signal of the genomic bin at position b (vertical red line). Moreover, the schematic distance function illustrates that CpGs close to position b will receive higher weights than CpGs located farther away.

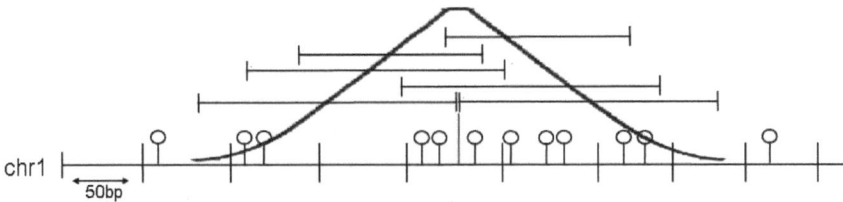

Figure 5: Schematic view of the genome vector created by defining a bin size of 50bp. In addition, CpGs are shown in a schematic way. A coupling factor is calculated for the centered genomic bin (marked by a red vertical line). For this, all CpGs within a maximal distance d are considered. The maximal distance d reflects the estimated average size of sequenced DNA fragments. There are several ways for calculating coupling factors. The simplest way is to count the number of CpGs in the surrounding of b but with a maximal distance of d. Alternatively, a weighting function can be applied to weight each CpG by its distance (dist) to the current genomic bin at position b.

Let C_{cb} be the coupling factor between a CpG at position c and a genomic bin at position b calculated based on an arbitrary weighting function and for any specified parameter d. Then $C_{tot} = \sum_c C_{cb}$ is the sum of coupling factors at the genomic bin b with respect to all CpGs at a genomic position c, where $|b-c| \le d$. For simplification, in the following, C_{tot} is called the coupling factor at a genomic bin b and gives a measure of local CpG density.

3.2.2. Calibration Curve

As we have created a genome vector that contains the raw signals at each genomic bin as well as an according coupling vector containing the calculated coupling factors at each genomic bin, the dependency of local MeDIP-seq signal intensities and local CpG densities can be examined. A dependency between CpG densities and MeDIP-seq signals can be made tangible by calculating the calibration curve [17]. Calculation of the calibration curve is achieved by first dividing the total range of coupling factors into regular levels. Second, all genomic bins are partitioned into these levels by considering their associated coupling factors. Finally, for each level of coupling factors, the mean signal and mean coupling factor of all genomic bins that fall into this level are calculated. As the calibration curve represents the averaged signals and coupling factors over the full range of coupling factors, it reveals the experiment specific dependency between signal intensities and CpG densities (see Fig. (**6**)).

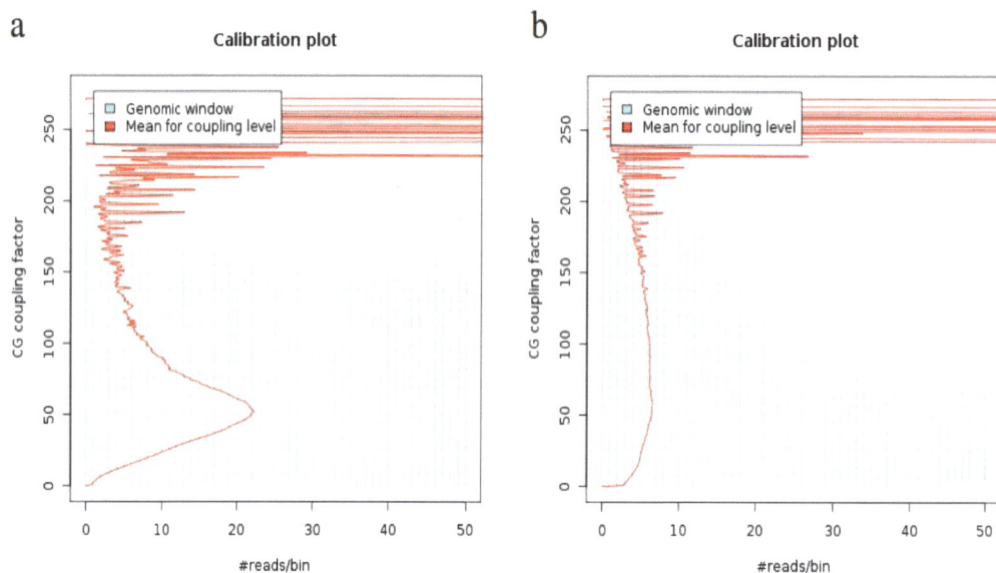

Figure 6: Calibration curves (red curve) for MeDIP-seq (**a**) and INPUT-seq (**b**) data from hESCs (1). Each data value within a calibration plot represents a genomic bin. The X-axis shows raw signals and the Y-axis shows the coupling factors for the genomic bins. The plots include genomic bins associated to MeDIP-seq signals ≤ 50 overlapping short reads per bin, only. The calibration curve (red curve) is very characteristic for MeDIP-seq experiments (see panel a). For low levels of CpG coupling factors, the calibration curve illustrates that MeDIP-seq signals, in average, increase just because of an increasing CpG density. An analogous dependency is, as expected, not observable for INPUT derived sequencing data (see panel b). The noise of the calibration curve in the high range of coupling factors results by a decrease of the number of genomic bins associated to high levels of coupling factors. Therefore, the calculated means for high levels of coupling factors are influenced by single outliers.

In fact, for the low range of coupling factors, the calibration curve indicates that the MeDIP-seq signals, in average, increase because of an increasing CpG density. Therefore, an increased signal is not necessarily caused by a higher level of mCpGs but scales with the general CpG density (see Fig. (**6a**)). For higher levels of CpG densities, the mean MeDIP-seq signals decrease. It has been shown [8, 16] that in mammalian cells, methylation is negatively correlated to CpG densities. In other words, regions of low CpG density tend to be high methylated, whereas regions of high CpG density tend to be mainly unmethylated. Therefore, it is assumed that the decrease of mean MeDIP-seq signals in higher levels of CpG densities is caused by the fact that in biological systems, regions of higher CpG denstities are mainly unmethylated. This circumstance implicates that the dependency between increased signal intensities caused by increased CpG densities is visible for regions of low CpG densities, only. For INPUT derived sequencing data (*i.e.,* sequencing sonicated but non immunoprecipitated genomic DNA fragments) the dependency of CpG density and sequencing signals is not observable (see Fig. (**6b**)). Therefore, the calibration plot is very characteristic for MeDIP-seq data and the quality of the enrichment step of the MeDIP experiment can be estimated by visual inspection of the progression of the calibration curve.

3.2.3. BATMAN Normalization

Based on the calibration curve, Down *et al.,* [17] developed a Baysian deconvolution strategy (BATMAN) where MeDIP signals observed at each genomic bin b depend on the methylation states of all nearby CpGs, weighted by the coupling factors between those CpGs and the genomic bin. Let m be a given set of methylation states, where m_c indicate the methylation state of a CpG at position c, and let B be a complete set of MeDIP-seq observations at the genomic bins b, then Down *et al.,* defined a probability distribution $f (B|m)$ which models MeDIP-seq observations given a set of methylation states. For this, they model the calibration curve by a polynomial model of order 2 and consider a rectified Gaussian error model. The resulting conditional probability distribution function f $(B|m)$ models MeDIP-seq data with respect to the coupling factor dependent calibration curve. By standard Bayesian inference, Down *et al.,* infer the posterior distribution $f (m|B)$ of the methylation state parameter given the MeDIP-seq data [17]. For 100bp windows, the BATMAN software finally generates quantitative methylation profile information. Down *et al.,* have shown that processing MeDIP-seq data with BATMAN significantly improves correlation of normalized methylation values to independently generated bisulfite sequencing data [8]. However, BATMAN was originally developed for analyzing MeDIP-chip data and its application to sequencing data is disproportional time consuming. Therefore, we have developed an alternative normalization method (MEDIPS, see section (**3.2.4.**)) that is comparable to BATMAN with respect to performance but outperforms computation time by orders of magnitude [1].

3.2.4. MEDIPS Normalization

3.2.4.1. Reads Per Million (rpm)

For each pre-defined genomic bin, the genome vector stores the number of provided overlapping extended short reads (these are the raw MeDIP-seq signals). Based on the total number of provided short reads (*n*), raw MeDIP-seq signals can be transformed into a reads per million (*rpm*) format in order to assure that coverage profiles derived from different biological samples are comparable, although generated from differing amounts of short reads. Let x_{bin_i} be the raw MeDIP-seq signal of the genomic bin *i*, where $i = 1,...,m$ and *m* is the total number of genomic bins, then the *rpm* value of the genomic bin is simply defined as:

$$rpm_{bin_i} = \frac{x_{bin_i} \cdot 10^6}{n}$$

3.2.4.2. Relative Methylation Score (rms)

After having calculated the genome vector from MeDIP-seq data (see section (**3.1.**)), MEDIPS calculates an according coupling vector (see section (**3.2.1.**)). Subsequently, analogous to BATMAN [17], MEDIPS investigates the dependency of MeDIP signals and local CpG density, by calculating the calibration curve (see section (**3.2.2.**)). The calibration curve reveals that, in average, an increase of MeDIP-seq signals is caused by an increasing CpG density. This approximately linear dependency is visible for the low range of coupling factors, only. For higher levels of CpG densities, mean MeDIP-seq signals decrease. As mentioned above, it is assumed that this decrease is caused by the fact that in mammalian cells, regions of higher CpG denstities are mainly unmethylated. Therefore, a continuing linear dependency of MeDIP-seq signals for higher levels of CpG densities is assumed. Analogous to Down *et al.,* [17], the local maximum of mean MeDIP-seq signals of the calibration curve in the lower part of coupling factors is identified. Let $y = y_1, ..., y_l$ be the mean coupling factors, and let $x = x_1, ..., x_l$ be the according mean MeDIP-seq signals of the calibration curve, where l is the number of tested coupling factor levels, and $i = 1, ..., l$, then the smallest level i is identified, where x_i is a local maximum. Let i_{max} be the according identified level of i, then $y_{max} = y_1, ..., y_{i_{max}}$ and $x_{max} = x_1, ..., x_{i_{max}}$ is the part of the calibration curve in the low range of coupling factors, where an approximately linear dependency between MeDIP-seq signals and coupling factors is observed (see Fig. (**7**)).

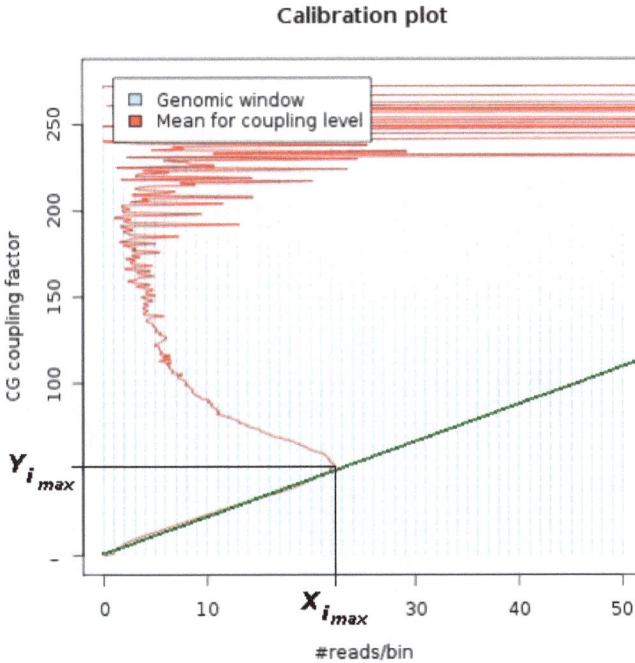

Figure 7: Calibration plot indicating the estimated linear dependency (green line) between MeDIP-seq signals (X-axis) and CpG density (Y-axis).

Here, x_{max} can be explained by a function of y_{max} as $x_{max} = f(y_{max}) + \varepsilon$, where ε is an error variable (*i.e.,* measurement errors) that is expected to spread by chance and therefore, its expectation value is $E(\varepsilon) = 0$. Because a linear dependency between x_{max} and y_{max} is assumed, x_{max} can be described as

$x_{\max} = \alpha + \beta \cdot y_{\max} + \varepsilon$, where the parameter α is the theoretical y-intercept, and the parameter β is the theoretical slope. Based on the pre-calculated $x_{\max}$ and $y_{\max}$ vectors, linear regression is performed, in order to identify a suitable linear model. Linear regression estimates concrete values a and b for the parameters α and β so that it is valid: $x_{\max_i} = a + b \cdot y_{\max_i} + e_i$, where $i = 1, ..., i_{\max}$. Here, the residuum e_i reflects the difference between the regression curve $a + b \cdot y_{\max_i}$ and the measurements for $x_{\max_i}$. Moreover, $x_{\max_i}$ can be replaced by an estimate $\hat{x}_{\max_i}$, where $x_{\max_i} - \hat{x}_{\max_i} = e_i$ and therefore, it is valid: $\hat{x}_{\max_i} = a + b \cdot y_{\max_i}$.

For estimating approximate values for the unknown parameters α and β, the least squares approach is utilized. In principle, the least squares approach identifies concrete regression coefficients a and b so that the sum of squared residues e_i, and therefore, the differences between the $x_{\max_i}$ variables of the calculated calibration curve and the $\hat{x}_{\max_i}$ variables of the resulting regression curve, are minimized. Subsequently, for the low range of coupling factors, the observed progression of the calibration curve can be modelled. As discussed above, a continuing linear dependency between MeDIP-seq signals and CpG density is expected for the higher range of coupling factors. Based on the obtained linear model parameters, concrete $\hat{x}_{\max_i}$ values can be calculated for the full range of coupling factors. Therefore, $\hat{x} = \hat{x}_1, ..., \hat{x}_{\max_i}, ..., \hat{x}_l$ are the estimated mean MeDIP-seq signals over the full range of coupling factor levels l, calculated with respect to the obtained linear model parameters (see green curve in Fig. (7)). In the following, the obtained $\hat{x}$ values are considered as the expected MeDIP-seq signals of genomic bins associated to the according coupling factor levels of i, where $i=1, ., l$ and l is the number of coupling factor levels. For MeDIP-seq data normalization, $\hat{x}$ is utilized in order to weight the observed MeDIP-seq signals of the genomic bins by the expected MeDIP-Seq signals. This $\dfrac{\text{observed}}{\text{expected}}$ ratio serves as a divergence indicator that allows for estimating the strength of MeDIP-seq signal enrichments with respect to local CpG densities. Let (x_{bin_i}, y_{bin_i}) be the raw MeDIP-seq signal of the genomic bin i (*i.e.,* the number of overlapping extended short reads), and the pre-calculated coupling factor at the genomic bin i, where $i = 1, ..., m$ and m is the total number of genomic bins, then the normalized relative methylation score is defined as $rms_{bin_i} = \log 2(\dfrac{x_{bin_i} \cdot 10^6}{\hat{x}_{bin_i} \cdot n})$, where $\hat{x}_{bin_i} = a + b \cdot y_{bin_i}$

is the estimated weighting parameter obtained by considering the coupling factor y_{bin_i} of the genomic bin i, and n is the total number of short reads considered for the generation of the genome vector. Based on the total number of short reads (n), the raw MeDIP-seq signals are, in parallel, transformed into a reads per million (*rpm*) format in order to assure that *rms* values are comparable between methylomes generated from differing amounts of short reads. We consider the *rms* values as the normalized MeDIP-seq signals corrected for the effect or varying efficiency of antibody binding dependent on local densities of methylated cytosines.

We have shown [1] that BATMAN [17] and MEDIPS massively improve correlation of MeDIP-seq [17] and bisulphite sequencing [8] data, and that results from both normalization methods can be well correlated.

4. FULL GENOME DNA METHYLATION ANALYSES OF HESCS

The most comprehensive DNA methylome of human embryonic stem cells at base resolution has been generated by whole genome shotgun bisulfite sequencing (WGSBS) and based on 1.16 billion reads that aligned uniquely to the human reference sequence [10]. Most interestingly, Lister *et al.,* discovered that almost 25% of all cytosines at which DNA methylation is identified were in non-CpG context. In contrast, differentiated human cells, like fetal lung fibroblasts or adult cell types, show only little or no non-CpG methylation [10, 30]. Although it remains difficult, if possible at all, to discriminate between CpG and non-CpG methylation using immunoprecipitation based approaches, MeDIP-seq covers nearly as many CpGs per sample genome as the more expensive WGSBS approach [14]. An advantage of the MeDIP approach is the generation of unbiased, cost-effective and full-genome methylation levels without the limitations associated with methylation-sensitive restriction enzymes. We have investigated the methylation profile of a hESC line by applying MeDIP-seq coupled with MEDIPS [1]. Our results show that for regions of interest, like for CpG islands or promoters, normalized MeDIP-seq data can be well correlated to WGSBS methylation data. When analyzing cell types known to express non-CpG methylation, like hESCs, MeDIP-seq data normalization may even be improved by calculating coupling factors as a weighted combination of CpG and non-CpG densities.

4.1. DNA Methylation at Promoters and Transcription Factor Binding Sites

It is supposed that DNA methylation in proximal promoter regions influences the expression of the according gene by interfering with binding of transcription factors [2]. For selected cell types, it has been demonstrated that MeDIP-seq derived normalized promoter methylation show negative correlation to gene expression [18]. Interestingly, in hESCs non CpG but not CpG promoter methylation shows negative correlation with gene expression [10]. The transcription factor OCT4 is known as a key regulator for maintaining pluripotency in the mammalian embryo [31, 32]. As an example, the OCT4 gene shows the presence of non-CG methylation in hESCs as well as cell-specific differential methylation when compared to differentiated cells like fetal lung fibroblasts [10] or definitive endoderm [1]. By ChIP-seq experiments, Lister er al. [10]identified binding events of the pluripotency associated transcription factors OCT4, NANOG, SOX2, and KLF4, as well as for the proteins TAF1 and p300. It was observed that relative cytosine methylation decreases at the identified interaction sites. Moreover, we have shown [1] that OCT4, KLF4, and TAF1 but not NANOG, SOX2, and p300 binding sites show bimodal CpG and methylation densities, suggesting distinct mechanisms for DNA methylation dependent regulation of transcription factor binding.

4.2. DNA Methylation Changes During Differentiation

Several studies have investigated DNA methylation changes during differentiation of hESCs [1, 10, 30, 33]. It has been shown that the degree of global DNA methylation was inversely correlated with differentiation status, where the highest level of methylation was seen in undifferentiated hESCs [30]. In general, the majority of methylation changes during differentiation account for de-methylation events [1]. Moreover, de- and *de novo* methylation events can be associated to functional known genomic regions. For example, a high proportion of methylation changes are observed at gene regions, including promoters, exons, and introns. Moreover, promoters with varying CpG densities show distinct patterns of methylation changes [1, 30]. Furthermore, differential methylation can be associated to CpG islands, transcription factor binding sites, gene activating histone modifications and to exon-intron bounderies [1, 10, 30]. Taken together, DNA methylation patterns are unique for different stages during differentiation of hESCs along the germ layers, and methylation alterations are associated with further known gene regulatory acting genomic regions.

CONFLICT OF INTEREST

None declared.

ACKNOWLEDGEMENT

None declared.

REFERENCES

[1]   Chavez L, Jozefczuk J, Grimm C, *et al.* Computational analysis of genome-wide DNA methylation during the differentiation of human embryonic stem cells along the endodermal lineage. Genome Res 2010.

[2]   Jaenisch R, Bird A. Epigenetic regulation of gene expression: how the genome integrates intrinsic and environmental signals. Nat Genet 2003; 33 Suppl: 245-54.

[3]   Meissner A, Mikkelsen TS, Gu H, *et al.* Genome-scale DNA methylation maps of pluripotent and differentiated cells. Nature 2008; 454 (7205):766-70.

[4]   Deng J, Shoemaker R, Xie B, *et al.* Targeted bisulfite sequencing reveals changes in DNA methylation associated with nuclear reprogramming. Nat Biotechnol 2009; 27 (4):353-60.

[5]   Chan EM, Ratanasirintrawoot S, Park IH, *et al.* Live cell imaging distinguishes bona fide human iPS cells from partially reprogrammed cells. Nat Biotechnol 2009.

[6]   Jones PA, Baylin SB. The epigenomics of cancer. Cell 2007; 128 (4):683-92.

[7]   Irizarry RA, Ladd-Acosta C, Wen B, *et al.* The human colon cancer methylome shows similar hypo- and hypermethylation at conserved tissue-specific CpG island shores. Nat Genet 2009; 41 (2):178-86.

[8]   Eckhardt F, Lewin J, Cortese R, *et al.* DNA methylation profiling of human chromosomes 6, 20 and 22. Nat Genet 2006; 38 (12):1378-85.

[9]   Rakyan VK, Down TA, Thorne NP, *et al.* An integrated resource for genome-wide identification and analysis of human tissue-specific differentially methylated regions (tDMRs). Genome Res 2008; 18 (9):1518-29.

[10]   Lister R, Pelizzola M, Dowen RH, *et al.* Human DNA methylomes at base resolution show widespread epigenomic differences. Nature 2009; 462 (7271):315-22.

[11]   Hayatsu H. Discovery of bisulfite-mediated cytosine conversion to uracil, the key reaction for DNA methylation analysis--a personal account. Proc Jpn Acad Ser B Phys Biol Sci 2008; 84 (8):321-30.

[12]   Wang RY, Gehrke CW, Ehrlich M. Comparison of bisulfite modification of 5-methyldeoxycytidine and deoxycytidine residues. Nucleic Acids Res 1980; 8 (20):4777-90.

[13]   Lister R, O'Malley RC, Tonti-Filippini J, *et al.* Highly integrated single-base resolution maps of the epigenome in Arabidopsis. Cell 2008; 133 (3):523-36.

[14]   Laird PW. Principles and challenges of genome-wide DNA methylation analysis. Nat Rev Genet 2010; 11 (3):191-203.

[15]   Meissner A, Gnirke A, Bell GW, *et al*. Reduced representation bisulfite sequencing for comparative high-resolution DNA methylation analysis. Nucleic Acids Res 2005; 33 (18):5868-77.

[16]   Weber M, Davies JJ, Wittig D, *et al*. Chromosome-wide and promoter-specific analyses identify sites of differential DNA methylation in normal and transformed human cells. Nat Genet 2005; 37 (8):853-62.

[17]   Bilgin CC, Bullough P, Plopper GE, Yener B. ECM-Aware cell-graph mining for bone tissue modeling and classification. Journal of Data Mining and Knowledge Discovery. 2009. 20(3): 416–438. doi: 10.1007/s10618-009-0153-2

[18]   Pelizzola M, Koga Y, Urban AE, *et al*. MEDME: an experimental and analytical methodology for the estimation of DNA methylation levels based on microarray derived MeDIP-enrichment. Genome Res 2008; 18 (10):1652-9.

[19]   Sanger F, Coulson AR. A rapid method for determining sequences in DNA by primed synthesis with DNA polymerase. J Mol Biol 1975; 94 (3):441-8.

[20]   Church GM. Genomes for all. Sci Am 2006; 294 (1):46-54.

[21]   Hall N. Advanced sequencing technologies and their wider impact in microbiology. J Exp Biol 2007; 210 (Pt 9):1518-25.

[22]   Schuster SC. Next-generation sequencing transforms today's biology. Nat Methods 2008; 5 (1):16-8.

[23]   Jacinto FV, Ballestar E, Esteller M. Methyl-DNA immunoprecipitation (MeDIP): hunting down the DNA methylome. Biotechniques 2008; 44 (1):35, 7, 9 passim.

[24]   Meehan RR, Lewis JD, Bird AP. Characterization of MeCP2, a vertebrate DNA binding protein with affinity for methylated DNA. Nucleic Acids Res 1992; 20 (19):5085-92.

[25]   Pomraning KR, Smith KM, Freitag M. Genome-wide high throughput analysis of DNA methylation in eukaryotes. Methods 2009; 47 (3):142-50.

[26]   Wilson IM, Davies JJ, Weber M, *et al*. Epigenomics: mapping the methylome. Cell Cycle 2006; 5 (2):155-8.

[27]   Zhang K, Martiny AC, Reppas NB, *et al*. Sequencing genomes from single cells by polymerase cloning. Nat Biotechnol 2006; 24 (6):680-6.

[28]   Li H, Ruan J, Durbin R. Map short DNA sequencing reads and calling variants using mapping quality scores. Genome Res 2008; 18 (11):1851-8.

[29]   Langmead B, Trapnell C, Pop M, Salzberg SL. Ultrafast and memory-efficient alignment of short DNA sequences to the human genome. Genome Biol 2009; 10 (3):R25.

[30]   Laurent L, Wong E, Li G, *et al*. Dynamic changes in the human methylome during differentiation. Genome Res 2010; 20 (3):320-31.

[31]   Nichols J, Zevnik B, Anastassiadis K, *et al*. Formation of pluripotent stem cells in the mammalian embryo depends on the POU transcription factor Oct4. Cell 1998; 95 (3):379-91.

[32]   Pesce M, Gross MK, Scholer HR. In line with our ancestors: Oct-4 and the mammalian germ. Bioessays 1998; 20 (9):722-32.

[33]   Brunner AL, Johnson DS, Kim SW, *et al*. Distinct DNA methylation patterns characterize differentiated human embryonic stem cells and developing human fetal liver. Genome Res 2009; 19 (6):1044-56.

CHAPTER 6

Transcriptional Co-Expression Analysis of Embryonic Stem Cells

Yu Sun and Ming Zhan[*]

National Institute of Aging, NIH, Baltimore, MD 20872

Abstract: The decision between differentiation and self-renewal of embryonic stem cells (ESCs) are controlled by a complex network formed by interacting genes or proteins. Identifying critical components of this regulatory network and exploring their behavior patterns is crucial towards understanding the underlying mechanisms controlling ESC differentiation and realizing their potentials in regenerative medicine. In this chapter, we describe the usage of co-expression analysis in identifying conserved and divergent genomic, transcriptomic, and network modules critical for the earliest stage of ESCs differentiation.

Keywords: Co-expression, cell cycle, transcriptome mapping, chromosomal domains, embryonic stem cell, embryonic bodies, comparative transcriptomics, transcriptional module, LIF, FGF, Nodal, Wnt, BMP, TGF-β, Oct4, Sox2, nanog, JAK/STAT, pluripotency.

INTRODUCTION

The microarray technology enables researchers to monitor the transcription levels of thousands of genes simultaneously. In the embryonic stem cell (ESC) research, microarray was initially used to generate static images of gene expression profiles in ESCs and their early-differentiated stage, embryonic bodies (EBs). Genes that showed differential expression between ESCs and EBs were identified through statistical analysis such as ANOVA or t-tests. While beginning to shed light on the mechanisms underlying ESCs pluripotency [1-5], these earlier studies often overlooked the concerted changes of genes expression behavior. Various studies have shown that genes that have highly correlated expression behavior form modules that carry out critical biological functions [6-9]. And the importance of examining co-expression in studying biological problems has been demonstrated

---

*Address correspondence to M. Zhan: Department of Systems Medicine and Bioengineering, The Methodist Hospital Research Institute, Houston, TX 77030, USA; Tel: (713) 441-8939; E-mail: mzhan@tmhs.org

Ming Zhan (Ed)

[10-13]. In this chapter, we describe our studies that explore gene co-expression at the pathway, global, and chromosome domain levels. We examined ESC-critical pathways by identifying conserved and divergent co-expression patterns [14]. Based on the co-expression behavior of the genes, we constructed global co-expression networks, and indentified hub genes which played critical roles in the development of ESCs [14]. We further revealed chromosomal domains showing correlated expression profiles of adjacent genes on the genome [15]. Through the co-expression profiling, we explored and identified molecular mechanisms that are fundamental or species-specific in controlling ESC development.

CO-EXPRESSION ANALYSIS OF PATHWAYS

A transcriptome dataset consisted of different cell lines of human and mouse ESCs and EBs (hESC, mESC, hEB, mEB) were either generated by us or collected from public database. The data sets contained 9 ESC and 9 corresponding 14-day differentiated EB samples from human and mouse origins, respectively: the mouse ESC and EB expression data were determined from V6.5 (GSE3231 of GEO database), R1 (GSE2972) and J1 (GSE3749) cell lines; the human ESC and EB data included expression profiles of H1 [4], HES2 (E-MEXP-303 of the ArrayExpress database), BG01, BG02 and BG03 cell lines [3, 16, 17]. After eliminating transcripts with low signal, we assembled a dataset containing 6, 573 human-mouse orthologous gene pairs selected based on the Affymetrix probe database. The gene expression data were normalized using the quartile method for the Illumina datasets and the RMA method for the Affymetrix datasets, and then transformed into log2 ratios of expression values over the mean value of all samples for each probe.

First, we focused on pathways critical for ESC development [18, 19] such as AKT/PTEN, Cell Cycle, JAK/STAT (including PI3K), TGFβ (including ACTIVIN/NODAL and BMP), and WNT pathways. We used two different computational algorithms: Generalized Singular Value Decomposition (GSVD) and Comparative Partition around Medoids (cPAM) to detect both conserved and divergent co-expression patterns in these ESC-critical pathways and transcription factors underlying the co-expression. These analyses revealed the fundamental and species-specific mechanisms regulating ESC pluripotency.

GSVD is a matrix decomposition algorithm that transforms the two gene expression datasets from the genes × arrays spaces to two reduced and diagonalized "eigengenes" × "eigenarray" spaces. The eigengenes are shared by both data sets. Each eigengene has a corresponding angular distance indicating the significance of this eigengene in one dataset relative to that in the other. Let the expression data of n orthologous genes in p samples (assume $n > p$) from two species be represented by matrices $\mathbf{M} = [\mathbf{m}_1, \mathbf{m}_2 \ldots \mathbf{m}_n]^T$ and $\mathbf{H} = [\mathbf{h}_1, \mathbf{h}_2 \ldots \mathbf{h}_n]^T$, respectively. $\mathbf{m}_i \in \mathfrak{R}^{1 \times p}$ and $\mathbf{h}_i \in \mathfrak{R}^{1 \times p}$ denote the data column vectors. GSVD is given by a pair of decompositions (Eq. 1):

$$\mathbf{M}_{n \times p} = \mathbf{U}_{n \times p}\mathbf{C}_{p \times p}\mathbf{T}_{p \times p}^{T}$$

$$\mathbf{H}_{n \times p} = \mathbf{V}_{n \times p}\mathbf{S}_{p \times p}\mathbf{T}_{p \times p}^{T} \tag{1}$$

where $\mathbf{C}$ and $\mathbf{S}$ are diagonal matrices with singular value elements ($c_1, c_2 \ldots c_p$) and ($s_1, s_2 \ldots s_p$), respectively, and satisfy $c_j^2 + s_j^2 \equiv 1$ ($j = 1, 2, \cdots p$). $\mathbf{U}$ and $\mathbf{V}$ are column-orthogonal matrices. $\mathbf{T}^T$, the tailing matrix, which relates the two datasets, is invertible but not orthogonal [20, 21]. The columns of matrix $\mathbf{T}$, $\mathbf{t}_1, \mathbf{t}_2 \ldots \mathbf{t}_p$, list the expression of p latent factors, the so-called "eigengenes", across p samples in both datasets simultaneously. The relative contribution of each eigengene to each dataset can be measured with the fraction of variance it captures in each dataset, calculated as the ratio of the square of the corresponding diagonal element in matrix $\mathbf{C}$ (or $\mathbf{S}$) with the sum, scaled with the length (inner product) of the corresponding eigengene vector (Eq. 2).

$$R_j^M = \frac{c_j^2 \|\mathbf{t}_j\|}{\sum_{l=1}^{p} c_l^2 \|\mathbf{t}_l\|} \quad \text{and} \quad R_j^H = \frac{s_j^2 \|\mathbf{t}_j\|}{\sum_{l=1}^{p} s_l^2 \|\mathbf{t}_l\|}, \quad j = 1, 2 \cdots p \tag{2}$$

When defining co-expression gene clusters, two projection matrices of the expression of n genes onto the p eigengenes are generated first (Eq. 3):

$$\mathbf{P}_{nxp}^M = \mathbf{M}_{n \times p}\mathbf{T}_{pxp} \quad \text{and} \quad \mathbf{P}_{nxp}^H = \mathbf{H}_{n \times p}\mathbf{T}_{pxp} \tag{3}$$

The gene clusters can then be identified based on the sorted projection values under each eigengene. Genes showing relatively high or low projection values than average under each eigengene are grouped together as co-expression clusters.

The conserved co-expression clusters are identified from the eigengene that has the smallest difference between singular values derived from two datasets. The difference between the two singular values of an eigengene is measured by the angular distance (Eq. 4):

$$\theta_j = \arctan\left(\frac{c_j}{s_j}\right) - \frac{\pi}{4} \quad j = 1, 2, \cdots p$$

$$(4)$$

An angular distance of zero indicates that an eigengene has equal contribution to both datasets and thus likely to be conserved. We first identify the eigengene j' that has the smallest angular distance θ_j and then rank genes by sorting their projection values of two data sets under eigengene j'. Subsequently, we choose the top 10% of all genes with relatively higher projection values of the two datasets under eigengene j' to form clusters C_1^M and C_1^H, and the bottom 10% of genes with relatively lower projection values to form clusters C_2^M and C_2^H. The conserved gene clusters C_1 that contains genes shared between C_1^M and C_1^H, and C_2 containing genes shared between C_2^M and C_2^H were then generated.

For example, for the cell-cycle process, we examined 356 human-mouse orthologous genes expression profiles in ESCs and EBs using GSVD analysis (Fig. **1A, 1B**). Fig. **1A** shows the 18 eigengenes, computed as a linear combination of genes, representing common features between two datasets. Among them, eigengene 3 showed the smallest difference between the two singular values that it was associated with (Fig. (**1B**)), indicating that this eigengene had near-equal contributions to the variances of human and mouse datasets. We then projected the human and mouse gene expression data onto the space of eigengene 3, and identified two conserved co-expression gene clusters, C1 and C2 for the cell-cycle process.

Another way to identify conserved co-expression patterns across species is to use the cPAM method, which uses the partition around medoids (PAM) algorithm to perform comparative gene clustering. Previous studies have shown that PAM is a clustering method robust to noise and outliers [22]. PAM partitions the dataset into K clusters, trying to minimize the total intra-cluster variance, calculated by the squared error function $v = \sum_{k=1}^{K} \sum_{x_k \in S_k} (x_k - \mu_k)^2$, where there are K clusters S_k (k = 1, 2, ..., K), with μ_k being the medoid point of all the points $x_k \in S_k$. First, the input points are partitioned into K initial sets, followed by calculating the medoid of each set. Then a new partition is constructed by associating each point with their closest medoid. The medoids are then re-calculated for new clusters, and the process repeats by alternative application of these two steps of clustering around mediods and re-calculation of medoids until the medoids no longer change. The cPAM method starts by first assigning one species as the primary species, and the genes are clustered according to their expression profiles in this species by PAM. The genes of the second species are then arranged together on the matrix according to the clusters identified in the primary species. The procedure repeats by assigning the second species as the primary species and clustered accordingly. Fig. (1C) illustrates the result from the cPAM analysis of the cell-cycle orthologous genes.

The expression profiles of cell-cycle genes formed two co-expression clusters in human and mouse ESCs (H1 and H2 in human, and M1 and M2 in mouse). In total, 44% or 58% of the genes showed co-expression in human or mouse cells (*i.e.*, members of clusters H1 and H2, or M1 and M2). The conserved co-expression clusters (O1 and O2) are formed by genes shared between H1 and M1, and H2 and M2, respectively (Fig. 1D). The O1 cluster is formed by 49 genes down-regulated in human and mouse ESCs with average $r = 0.523$ in human and 0.717 in mouse cells, respectively. The O2 cluster is formed by 81 genes up-regulated in human and mouse ESCs with average $r = 0.747$ in human and 0.851 in mouse cells, respectively (Table 1). Together, they account to 37% of the total cell- cycle orthologous genes. The co-expression level in each of these gene clusters is statistically significant than what you would expect from random samples (P value<0.01 by random shuffling). The conserved co-expression clusters identified by cPAM and GSVD were largely similar, with O1 corresponding to the C1 cluster, and O2 to the C2 cluster.

We mapped the co-expression patterns we observed onto the core network of the cell-cycle to further examine how the transcriptional modulation is involved in the core activities of the cell-cycle in ESCs. As illustrated in Fig. 1E, many genes carrying out critical functions in cell-cycle regulation of ESCs showed conserved co-expression. For example, CCNA2 and CCNB2, cyclins of S and M phases, were members of O2 and up-regulated in both human and mouse ESCs. CCND2

Cell cycle

A

B

C

D

E

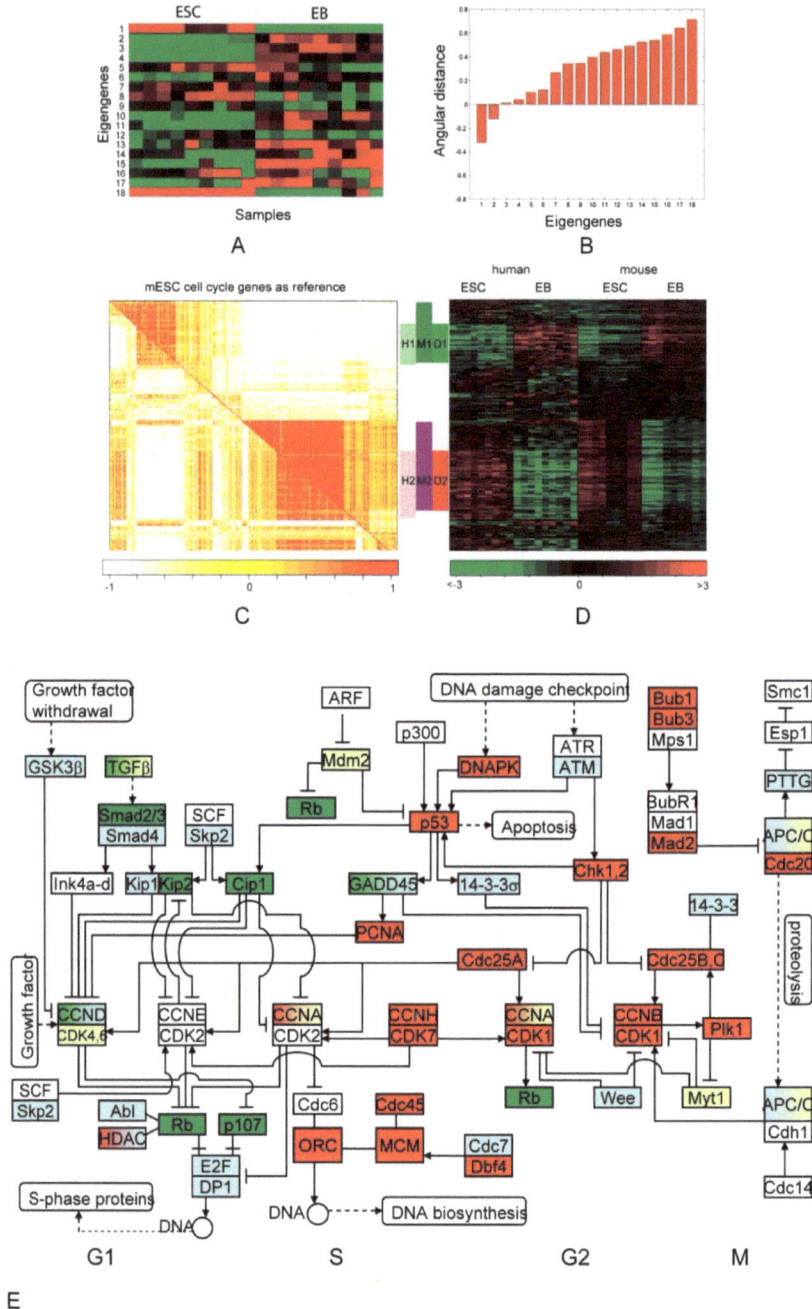

Figure 1: Identification of conserved co-expression clusters in the cell-cycle pathway of human and mouse ESCs using GSVD and cPAM. A) Heatmap of the tailing matrix T shared by human and mouse ESCs expression profiles by GSVD. The expression level of each eigengene

(rows) across all ESC and EB samples of human and mouse datasets (columns) is illustrated by the red-to-green color gradient representing high-to-low values. The matrix was normalized so that all rows had the length of 1. B) Bar plot of the angular distance between singular values, which shows the difference of eigengene's contribution to the total variances of human and mouse datasets. Eigengene #3 showed the smallest angular distance, indicating near-equal contributions to both datasets. It was then used to derive conserved gene clusters across species. C) Correlation matrix of gene expression profiles in human cells (the lower-left part) and mouse cells (the upper-right part), generated by cPAM. The light-to-intense color gradient on the graph represents low-to-high pair-wise correlation coefficient values between genes. The orders of the genes on the horizontal and vertical axes are the same and were determined by PAM. The genes were clustered into co-expression clusters H1 and H2 in human and M1 and M2 in mouse. Genes overlapped between H1 and M1 formed the conserved co-expression cluster O1. Genes shared between H2 and M2 formed O2. D) Heatmap of the normalized expression levels of the genes presented in the correlation matrix. The genes are listed in the same order as in C. The green-to-red color gradient represents low-to-high expression levels of a gene in comparison to the mean expression level across all samples. Genes in H1, M1 and O1 clusters were down-regulated, whereas the genes in H2, M2 and O2 were up-regulated in undifferentiated human and mouse ESCs. E) Core network of the cell-cycle pathway. The genes showing co-expression in both human and mouse cells are colored red or green (representing up- or down-regulation in ESCs). The genes showing co-expression in only one species are colored blue. The genes showing no co-expression at all are colored yellow. (adopted from [14]).

and CCND3 (cyclins of the G1 phase), the retinoblastoma protein (RB) and P107 (RBL1) were members of O1 and down-regulated in both ESCs. It has been shown that the cell-cycle regulation in mESCs relies on constitutively active CCNA: CDK2 and CCNE:CDK2 complexes rather than an active INK4A/CCND:CDK4/RB:E2F pathway [23, 24]. Our results are consistent with these experimental observations and further indicate that these mechanisms are possibly conserved among ESCs of different species. In addition, the up-regulation of S phase cyclins and down-regulation of G1 phase cyclins highlights the fact that ESCs have a shortened G1 phase and are primed for a rapid cell proliferation [25]. Other members of the up-regulated and conserved co-expression cluster O2 include subunits of the origin recognition complex (ORC1, ORC2, ORC3, ORC5 and ORC6), genes encoding mini-chromosome maintenance deficient proteins (MCM2, MCM3, MCM4, MCM5, MCM6 and MCM7), DNA replication initiation factor (CDC45), and replication proteins (RPA1, RPA2 and RPA3). These results suggest that both human and mouse ESCs have an elevated and tightly regulated DNA replication activity and a shortened cell-cycle, a conclusion also supported by previous experimental observations [24, 25]. In contrast to members of the core cell-cycle network, many cell-cycle related genes outside of the core network showed a divergent co-expression pattern. For example, 34 genes of the M2 cluster showed co-expression in mouse but not human ESCs, including CDC elements (CDC23, CDC37 and CDC37l1), cell-cycle regulation factors (CCNC, CCND1), and cell-cycle phase related factors (DP1, E2F4 and SKP2). Fifteen genes of the H2 cluster showed co-expression in human but not mouse ESCs, including M-phase factors BUB1, KATNB1, MKI67, NCAPD2, SMAC4I1 and WEE1. In summary, our results suggest that tightly coordinated transcriptional modulation is an essential mechanism for the core activities of the cell-cycle during early differentiation of human and mouse ESCs. However, these conserved core activities may be regulated differently in human and mouse ESCs by species-specific mechanisms outside the core network.

The co-expression clusters identified by cPAM and GSVD analyses of other ESC-critical pathways are summarized in Table **1**. The co-expression in each gene cluster was statistically significant (P value<0.001). The conserved co-expression clusters identified by GSVD and cPAM methods were largely consistent.

The canonical WNT signaling pathway is critical in maintaining pluripotency in both human and mouse ESCs [26, 27]. Our analysis showed that the conserved co-expression mostly occurred among downstream target genes, while key components of this pathway had divergent co-expression patterns. Among the WNT ligand genes, only WNT5a showed conserved co-expression across species. WNT1 and WNT5b showed co-expression in human but not mouse ESCs, whereas WNT2, WNT2b and WNT6 showed co-expression in mouse but not human ESCs. WNT3 and WNT10b showed no co-expression in either human or mouse ESCs. Of the signal transducing scaffold factors, DVL3 was co-expressed only in hESCs, whereas DVL2 was co-expressed only in mouse ESCs. Similarly, AXIN1, FRAT2, CSNK1E, CTNNB1 and GSK3b all showed species-specific co-expression behavior, with AXIN1 and FRAT2 co-expressed in hESCs only, while CSNK1E, CTNNB1 and GSK3b co-expressed only in mouse ESCs. The detailed mechanism of WNT signaling in regulating ESC pluripotency remains unclear [26-29]. Studies of the effect of the inhibition of G3K3b in human and mouse ESCs [26] or PP2A in mESCs [27] indicates that the activation of the canonical

Table 1: Summary of co-expression gene clusters identified in ESC-critical pathways by GSVD and cPAM. (adopted from [14]).

Pathway	No. of genes examined	Co-expression cluster	No. of gene in cluster	Average r	Expression in undifferentiated ESCs
Cell Cycle	356	H1	60	0.473	Down
		H2	96	0.747	Up
		M1	84	0.676	Down
		M2	123	0.837	Up
		O1	49	0.523 (H) 0.717 (M)	Down
		O2	81	0.747 (H) 0.851 (M)	Up
WNT	92	H1	17	0.645	Up
		H2	19	0.59	Down
		M1	25	0.558	Up
		M2	40	0.607	Down
		O1	10	0.713 (H) 0.600 (M)	Up
		O2	14	0.596 (H) 0.715 (M)	Down
JAK-STAT	58	H1	19	0.581	Down
		H2	9	0.519	Up
		M1	22	0.658	Down
		M2	12	0.663	Up
		O1	15	0.610 (H) 0.674 (M)	Down
		O2	7	0.691 (H) 0.746 (M)	Up
TGFβ	54	H1	21	0.534	Down
		H2	3	0.861	Up
		M1	30	0.562	Down
		M2	16	0.49	Up
		O1	17	0.520 (H) 0.611 (M)	Down
		O2	3	0.861 (H) 0.915 (M)	Up
AKT/PTEN	63	H1	13	0.539	Up
		H2	18	0.601	Down
		M1	16	0.483	Up
		M2	23	0.555	Down
		O1	7	0.597 (H) 0.550 (M)	Up
		O2	12	0.637 (H) 0.675 (M)	Down

WNT pathway promotes ESC self-renewal. On the other hand, direct activation of the WNT signaling using a recombinant WNT3a protein shows that WNT activation promotes both differentiation and self-renewal in hESCs [28, 29]. Taken together, the experimental observations and our computational analyses all suggest that the WNT canonical pathway plays a complex role in human and mouse ESCs, and the functions and mechanisms of WNT signaling are not evolutionarily conserved. However, our results suggest that WNT5a may be important to ESC development as a part of conserved modulation of gene expression in canonical WNT signaling and merits further investigation.

Of the 58 human-mouse orthologous genes genes in JAK/STAT and PI3K pathways we studied, 28 were co-expressed in human ESCs, and 44 were co-expressed in mouse ESCs. Of them, 22 genes formed the conserved co-expression clusters O1 and O2. The conserved co-expression cluster O1 contains JAK1, P40/ISGF3G, STAT2, certain receptors (CNTFR, GHG, IFNGR and IL10RB) and downstream target genes (BCL2L1, CCND2 and CCND3), as well as key PI3K pathway elements (AKT1, PIK3R1, PIK3CD and PIK3R4). This observation indicates that JAK-mediated signaling through activation of STAT2 and PI3K plays a critical role to ESC differentiation across species. Interestingly, our analysis showed that STAT3 was tightly co-expressed with JAK1 and other genes in mouse ESCs only, highlighting the fact that LIF signaling by STAT3 activation through the JAK/STAT cascade is essential in mouse ESCs, but not required in human ESCs [30, 31]. Previous studies have shown that the PI3K pathway can be activated by LIF or insulin signaling and inhibits the activity of GSK3β, a WNT signaling inhibitor, in mESCs [32, 33]. Our observation of the conserved co-expression among PI3K pathway members (AKT1, PIK3R1, PIK3CD and PIK3R4) suggests that transcriptional modulation in the PI3K pathway may be essential for both human and mouse ESCs.

The TGFβ network we examined includes members of BMP and ACTIVIN/NODAL pathways. Nineteen of the 54 orthologous genes showed conserved co-expression behavior between human and mouse ESCs. Most of the conserved co-expression genes are members of the ACTIVIN/NODAL pathway. Key pathway components such as signal transducers SMAD2 and SMAD3, the target and repressor gene FST and the target gene PITX2 were members of the conserved cluster O1, while the co-receptor TDGF1 and antagonist LEFTB belonged to O2. The conserved co-expression pattern suggests that tight transcriptional regulation is an essential mechanism for ACTIVIN/NODAL-directed signaling. Previous studies indicated a critical role of ACTIVIN/NODAL signaling through SMAD2/3 in maintaining the pluripotency of human ESCs [34-36]. Our results suggest and that this pathway maybe fundamental in both human and mouse ESCs. In contrast, members of the BMP pathway showed a divergent co-expression pattern across species. Receptors BMPR1A and BMPR2, signal transducers SMAD1, SMAD4 and SMAD5, and the target gene ID1 were co-expressed in mouse ESCs only. This divergent co-expression pattern suggests distinct roles played by BMP signaling in human and mouse ESCs, an observation consistent with previous experimental results. In mESCs, active BMP signaling through SMAD1/5 is reported to collaborate with LIF signaling to maintain pluripotency [37]. In hESCs, however, BMP signaling needs to be suppressed in order to maintain pluripotency since its activation will induce hESCs differentiation along the trophectoderm path [38].

Of the 63 orthologous genes of the AKT/PTEN pathway we studied, 31 were co-expressed in hESCs, 39 in mESCs, and 19 of them formed conserved co-expression clusters. In particular, key pathway components such as PIK3R4, phospholipase C (PLCB3 and PLCG2), the signal transducer PDK1, and the downstream target gene FOXO1 belonged to the conserved up-regulated cluster O2 in ESCs. While PIK4CA, the signal transducer AKT1, kinases PIP5K1C, TESK1 and ITPKB, phosphatases INPPL1, PLCD1 and INPP5A, and downstream target genes CCND2 and CCND3 belonged to the conserved down-regulated cluster O1 in ESCs. The observation that the transcription of these key pathway components, along with downstream target genes, are tightly regulated across species, supports the idea that the AKT/PTEN signaling plays a critical role in regulating ESC development.

Genes that form co-expression clusters are likely to be regulated through common transcriptional regulatory mechanisms [39]. We further sought to uncover shared transcription factors underlying the conserved co-expression patterns in the ESC-critical pathways using promoter sequence analysis. We retrieved 7 kb promoter sequences for each gene in the conserved co-expression clusters from the

Promoser database (biowulf.bu.edu/zlab). We then identified potential transcription factor binding sites by scanning the promoter sequences against the Transfac 9.0 database using the software Match (www.gene-regulation.com), with matrix similarity threshold equals 0.8 and core similarity equals 0.85. The binding sites of the transcription factor NANOG were detected based on the consensus sequence (C/G) (G/A) (C/G)C (G/C)ATTAN (G/C) [40]. The EBM was recognized based on the consensus sequence A/CGGAA/T [41] and the TLBM was based on CTTTGA/TA/T [42]. The FGF response element (FRE) was detected by scanning ETS binding motifs (EBM) and TCF/LEF binding motifs (TLBM), which lie adjacent to each other on the same strand.

We listed in Table **2** the transcription factors whose binding sites were discovered in more than 50% of human-mouse orthologous genes and are statistically over-represented in at least one of the conserved co-expression clusters we discovered in the ESC-critical pathways (P value<0.01 by Fisher's exact test). Evolutionarily conserved in both their genomic sequences and the transcriptional activities of their target genes, these transcription factors probably carry out critical functions in regulating the expression of these co-expression genes. Some of the transcription factors were present in most of the pathways and may represent common regulators of conserved transcriptional co-expression among different pathways, while others seemed to be pathway specific as they were found in only one or two pathways (Table **2**). Some of the "conserved" and "common" transcription factors have been implicated to be critical in regulating ESC self-renewal or differentiation. For example, OCT4 and NANOG form the central regulatory circuitry together with SOX2 in ESCs [43, 44]. ETS and TCF/LEF are effectors of ESC-critical canonical WNT and FGF/RAS/MAPK pathways [19, 45, 46]. One interesting observation is that the binding sites of ETS and TCF/LEF were often found to be adjacent to each other on the promoter sequences of co-expressed genes in these pathways and form the so called FGF response element (FRE) [47]. Being associated with most human and mouse genes in all conserved co-expression clusters in the ESC-critical pathways we studied, the FRE appears to be evolutionary conserved and hence fundamental in regulating transcriptional activities critical to ESC development. FREs are also present on the regulatory region of SOX2 [48]. Moreover, the cognate motif of SOX2 (CA/TTTGTT) is similar to that of TCF/LEF (CTTTGA/TA/T). This raises the possibility that SOX2 competes with TCF/LEF on the FRE in cooperation with ETS proteins, or TCF/LEF competes with SOX2 on the target genes of the SOX2/OCT4/NANOG co-binding. Taken together, FGF activating the FRE may integrate with AKT/PTEN, Cell Cycle, JAK/STAT (including PI3K), TGFβ (including ACTIVIN/NODAL and BMP), and WNT pathways, as well as the central regulatory circuitry formed by SOX2, OCT4 and NANOG, in determining the fate of ESCs.

Table 2: Transcription factors which showed binding sites among most genes of both human and mouse and statistically over-represented in conserved co-expression clusters. (adopted from [14]).

	Conserved Transcription Factors
Present in all 5 pathways*	OCT1; OCT4; CDP; CDXA; ETS; GATA4; MRF2; NCX; NKX2-5; POU3F2; SPZ1; SREBP1; SRY; TBX5; TST1; TTF1; VDR; ZF5; ZIC2; AP2; DBP; MAF; MAZ; MYB; RFX; SP3; TCF11; HAND1; E47; HNF-3ALPHA; HNF-4ALPHA; PAX; SREBP; FOX; LEF1TCF1; STAT; TEL2; NANOG
Present in only 1-2 pathways	ETF; RUSH-1ALPHA; SF1; SOX9; AR; CART1; COUP; IPF1; KROX; LXR; LYF1; MYOD; NF1; PEBP; SMAD3; SZF1-1

*The 5 pathways include AKT/PTEN, Cell Cycle, JAK/STAT (including PI3K), TGFβ (including ACTIVIN/NODAL and BMP), and WNT.

GLOBAL CO-EXPRESSION NETWORK OF ESCs

Having identified conserved co-expression behaviors in ESC-critical pathways, we sought to study the co-expression patterns in hESCs and mESCs on a global scale by constructing global co-expression networks related to the early differentiation of human and mouse ESCs. A co-expression network consists of nodes representing genes and links representing co-expression relationship between the genes. The links on the co-expression networks were determined by pair-wise correlation coefficients (r) between the expression profiles of genes. If r was above a threshold value T_r, then there will a link between the pair of genes. T_r is determined according to the scale-free criterion, which is measured by the square of the correlation coefficient (R^2) between log ($P(k)$) and log (k), where k denotes the connectivity of a node (*i.e.,* the number of links connecting a node to other nodes). $P(k)$ gives the probability that a selected node has exactly k links, which is calculated as the number of the nodes at a given k value divided by the total number of nodes. At or above the threshold values, genes were considered to be co-expressed and the derived networks obeyed a power law distribution and were scale-free. The scale-free property, a common feature among biological networks [9, 49-51], suggests that the topology of the co-expression networks is dominated by a few highly-connected genes, the so-called hub genes, which form the core of the network and link the other less-connected genes to the system. Using this method, we constructed the co-expression networks of hESCs and mESCs, which consist of 6, 118 and 4, 120 genes, respectively. Based on the overlapping or shared components between the hESC and mESC co-expression networks, we further constructed a hESC-mESC conserved co-expression network, which consisted of 5, 712 genes. Both the hESC and mESC networks, as well as the hESC-mESC conserved network were scale-free.

Through an assessment of functional similarities between co-expressed genes in the ESC co-expression networks, we found that these co-expressed genes were significantly associated with biological functions than randomly selected gene pairs. In the assessment, a functional similarity score of a pair of genes A and B is measured by the number of GO terms shared between them, $(GO_A \cap GO_B)$, where GO_A or GO_B denotes the set of GO terms for genes A or B. The functional similarity scores were calculated for all the co-expressed gene pairs in the co-expression network, as well as randomly selected gene pairs. The difference between the resulting cumulative distributions of functional similarity scores of genes in the co-expression network and random gene pairs are examined by the Kolmogorov-Smirnov test for the statistical significance. Interestingly, the hESC-mESC conserved network showed a stronger functional similarity than either the hESC or the mESC networks, suggesting that natural selection favors the co-expressed gene pairs that are functionally coupled over those with less functional relevance in the networks.

The hub genes on biological networks are often associated with critical biological functions. Knocking out of such genes can be lethal [52]. Interestingly, all the hub genes of the co-expression networks we constructed showed significant expression activity changes between ESCs and EBs (multiple testing adjusted P value≤0.05). In addition, many of the hub genes were related to the cell-cycle, development, and signaling pathways critical for ESC. For example, RUVBL1, which is evolutionarily conserved and essential for viability of yeast, flies and worms [53], is critical for c-MYC and WNT signaling pathways [54-56], and for apoptosis, DNA repair and chromatin remodeling [57, 58]. ORC1L and ORC2L are important for DNA replication initiation. NASP, an H1 histone binding protein, may regulate early events of spermatogenesis [59]. JARID2 is involved in mouse embryogenesis by showing embryogenesis-specific expression, and participates in the negative regulation of cell proliferation signaling [60, 61]. Some hub genes, such as IGF2, LCK and PHC1, have been implicated in ESC development regulation by previous studies. PHC1, for example, is located adjacent to genes that are hallmarks of stem cells on human chromosome 12, including CD9, GLUT3, GDF3, NANOG, STELLA and SCNN1A [62]. The target genes of PHC1 need to be transcriptionally repressed in ESCs in order to maintain pluripotency. Activation of these genes would initiate ESC differentiation [63, 64]. LCK is one of the SRC genes, which are implicated in maintaining pluripotency in mESCs [65, 66]. IGF2, an insulin growth factor, shows aberrant genomic imprinting, abnormal hypermethylation and an altered epigenetic status in ESCs [67, 68]. Among transcription factors that are hub genes, HAND1, HMGA1, MYCN and OCT4 are known to be critical for ESC development [43, 69-71]. HOPX, NFYB, TGIF1, TFAM, TRIM28 and others, on the other hand, have unknown functions in ESCs and may serve as new candidates in future studies.

CHROMOSOMAL TRANSCRIPTOME MAP

We also investigated the gene co-expression and co-regulation behavior operating at the local chromosome level by constructing chromosomal transcriptome maps. We constructed the maps for hESC and hEB based on microarray expression datasets from hESC cell lines I6, BG01V, BG01, BG02 and BG03 [72] and their corresponding 14-day differentiated hEBs, determined by high density Illumina BeadArray [15]. First, we calculated the "gene neighbor co-expression index" for each gene, which is the average of pair-wise correlation coefficient values between the expression level of a gene and each of its upstream and downstream neighboring genes within a certain genomic window. To determine the size of the window (*i.e.,* number of neighboring genes) on the chromosome, we compared the mean gene neighbor co-expression index of different window sizes. When the window size increased from 2 to 20, the mean gene neighbor co-expression index decreased significantly. However, beyond 20 neighboring genes, the decrease of the co-expression index became less significant. Therefore, we used window size of 20 genes (10 upstream and 10 downstream) to calculate the gene neighbor co-expression index. We then conducted Monte-Carlo simulation to access the significance of the co-expression index. In the simulation, 10, 000 random data sets were generated by random shuffling position indices of the same number of genes on each chromosome. The resulting distributions of the co-expression index and mean co-expression index were fitted to the Gaussian density function. The P values of the co-expression index and mean co-expression index from the real dataset were then determined according to the simulated probability distribution.

Based on the simulated distribution, we used the co-expression index value of 0.3 as the cut-off for identifying chromosome domains of co-expression. This co-expression index value corresponded to the P value of 0.0004 in the hESC expression set and 0.01 in hEB. At this threshold, we identified 205 chromosome domains of co-expression in hESC and 549 co-expression domains in hEB. Among the co-expression domains, only 18 domains were shared between hESC and hEB. The remaining domains showed differential co-expression between hESC and hEB samples. This big difference in the co-expression patterns we observed indicates that upon hESC differentiation, there is considerable change in the co-expression behavior among neighboring genes. Fig. (**2**) shows the distribution of the co-expression domains on the genome. Through Fisher's exact tests, we found that the co-expression domains and genes within these domains were highly frequent on chromosomes 8, 11, 16, 17, 19, and Y in hESC, whereas in hEB, they are enriched on chromosomes 6, 11, 17, 19 and 20.

By assessing biological relevance of the co-expression domains using Fisher exact tests, we found that many of the domains were significantly enriched in genes with transcription regulation and chromosome organization, as well as genes with ESC-critical functions such as apoptosis, pattern specification, embryogenesis and morphogenesis, histogenesis and organogenesis. Some genes known to be critical for hESC pluripotency and self-renewal showed significant degree of co-expression with neighboring genes. For example, STAT3, which is a transcription factor critical in hESC self-renewal pathways and feed-back loops, was co-expressed with neighboring genes in hESC (co-expression index = 0.37) but not in hEB (co-expression index = -0.17). Other hESC-signature genes (*e.g.,* UTF1, TLE1, and OCT3/4) showed higher co-expression index in hEB (0.299, 0.285, and 0.23, respectively) than in ES (0.125, 0.08, and 0.10, respectively). However, many other hESC-signature genes did not show significant co-expression with neighboring genes in either hESC or hEB samples (Table **3**). The chromosome transcriptome maps thus illustrate how genome organization affects gene expression in human ESCs and help to identify new mechanisms and pathways controlling ESC self-renewal or differentiation.

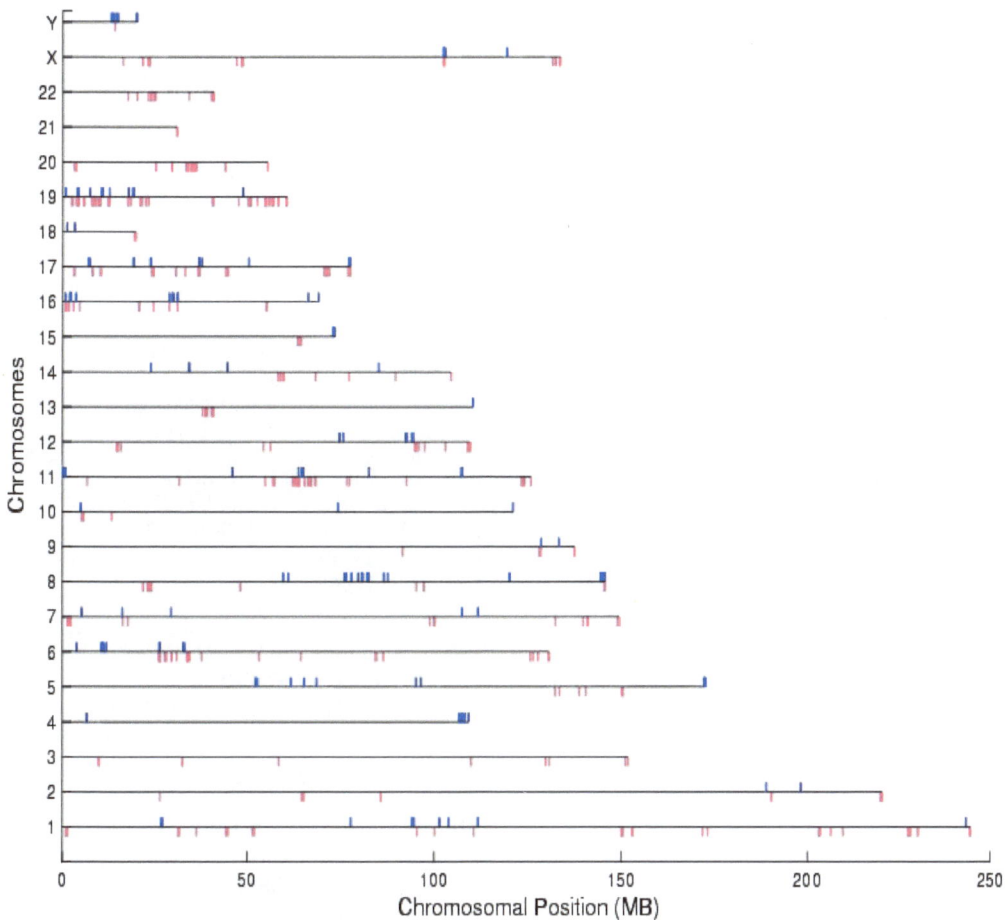

Figure 2: Genomic overview of co-expression chromosomal domains in human ES and EB cells. The domains were identified by the genes which have significant correlation in the expression profile with neighboring genes (10 upstream and 10 downstream) along the chromosome, at a co-expression index cut-off of 0.3. Blue bars on the chromosome represent co-expression domains in hESC, while red bars represent co-expression domains in hEB. (Adopted from [15]).

Table 3: ESC-signature genes and their co-expression with the neighboring genes. (Adopted from[15]).

Gene	Chromosome Location	Co-Expression Index (ES)	Co-Expression Index (EB)	Gene in Domain* (20 Neighboring genes: 10 upstream, 10 downstream)
FOXD3	1p32-p31	0.069	-0.083	INADL; FLJ10884; LOC163782; USP1; DOCK7; ANGPTL3; 400756; AUTL1; 199897; LOC199899; FOXD3; ALG6; ITGB3BP; PGM1; ROR1; 219612; MGC35130; KRTAP4–7; KIAA1573; KIAA1579; JAK1
TDGF1	3p21.31	-0.157	-0.016	401062; XCR1; CCR1; CCR3; CCR2; CCR5; CCRL2; LTF; TMEM7; LRRC2; TDGF1; FLJ36525; TMIE; TSP50; TESSP5; TESSP2; MYL3; PTHR1; MGC23918; HYPB; KIF9
SOX2	3q26.3–q27	0	0	GNB4; BAF53A; MRPL47; 133993; NDUFB5; USP13; PEX5R; TTC14; FXR1; LOC131118; SOX2; 401103; 402152; LOC142678; ATP11B; RP42; MCCC1; LAMP3; KIAA0861; B3GNT5; KLHL6
ABCG2	4q22	-0.039	-0.029	DHRS8; NUDT9; SPARCL1; DSPP; DMP1; LOC153218; IBSP; MEPE; SPP1; PKD2; ABCG2; DKFZp761G058; CEB1; MGC14156; DRLM; TIGD2; LOC285513; SNCA; MMRN; IRAK1BP1; TMSL3
LIFR	5p13-p12	0.187	-0.042	FLJ30596; FLJ25422; SLC1A3; IDN3; FLJ13231; NUP155; FLJ10233; GDNF; 147975; FLJ39155; LIFR; 253254; 401182; OSMR; MGC39830; FYB; C9; DAB2; PTGER4; OSRF; PRKAA1
IL6ST	5q11	0	0	GZMA; FLJ37927; 345643; UNG2; DHX29; KIAA0052; PPAP2A; FLJ90709; DDX4; CRL3; IL6ST; FLJ11795; 345645; MGC33648; FLJ35954; DKFZp761C169; 345651; SNK; FLJ33641; RAB3C; PDE4D
OCT3/4	6p21.31	0.102	0.23	IER3; DDR1; 389376; DPCR1; C6orf15; PSORS1C1; CDSN; PSORS1C2; C6orf18; TCF19; OCT3/4; LOC253018; HLA-C; HLA-B; MICA; HCP5; MICB; BAT1; ATP6V1G2; NFKBIL1; LTA
CER1	9p23-p22	-0.003	0.023	NIRF; GLDC; GASC1; PTPRD; TYRP1; 286343; MPDZ; 401492; NFIB; ZDHHC21; CER1; FLJ25461; C9orf52; SNAPC3; PSIP2; FLJ39267; C9orf39; SH3GL2; ADAMTSL1; FLJ35283; MGC35182
TLE1	9q21.32	0.084	0.285	PCSK5; FLJ11149; GCNT1; C9orf65; CHAC; GNA14; GNAQ; FLJ12643; PSAT1; TLE4; TLE1; FLJ43950; 389763; FLJ31614; MGC20553; UBQLN1; GKAP42; KIF27; C9orf64; HNRPK; C9orf76

Table 3: cont….

NODAL	10q22.1	-0.001	-0.161	C10orf35; COL13A1; H2AFY2; AMID; MGC34695; SARA1; PP; OT7T022; FLJ10751; EIF4EBP2; NODAL; KIAA1274; PRF1; ADAMTS14; C10orf27; 338611; SGPL1; PCBD; UNC5B; SLC29A3; CDH23
UTF1	10q26	0.125	0.298	C10orf39; DPYSL4; PKE; LOC170394; LOC170393; INPP5A; NKX6-2; FLJ25954; GPR123; KIAA1768; UTF1; VENTX2; ADAM8; TUBGCP2; ZNF511; CALCYON; UPA; FLJ26016; ECHS1; PAOX; LOC92170
FGF4	11q13.3	-0.057	0	CPT1A; MRPL21; IGHMBP2; MRGD; MGC21621; TPCN2; MYEOV; CCND1; ORAOV1; FGF19; FGF4; FGF3; 399920; ORAOV2; FADD; PPFIA1; EMS1; SHANK2; 399921; LOC220070; DHCR7
NANOG	12p13.31	-0.06	0.007	RBP5; CLSTN3; PXR1; 341392; M160; CD163; APOBEC1; GDF3; DPPA3; CLECSF11; NANOG; SLC2A14; SLC2A3; FHX; C3AR1; DKFZP566B183; CLECSF6; FLJ10408; CLECSF8; CLECSF9; AICDA
STAT3	17q21	0.37	-0.171	201181; LGP2; GCN5L2; HspB9; RAB5C; KCNH4; HCRT; LGP1; STAT5B; STAT5A; STAT3; PTRF; ATP6V0A1; NAGLU; HSD17B1; DPCK; TCFL4; HUMGT198A; LOC162427; TUBG1; TUBG2
ERAS	Xp11.23	-0.036	0.171	SLC38A5; FTSJ1; PPN; EBP; RBM3; WDR13; WAS; SUV39H1; GATA1; HDAC6; ERAS; PCSK1N; TIMM17B; PQBP1; SLC35A2; PIM2; DKFZp761A052; KCND1; TFE3; JM11; JM4

*The genes in domains are arranged as they are on the chromosome from 5' to 3'.

SUMMARY

We examined transcriptional co-expression guiding ESC development in the framework of ESC-critical pathways, global networks and chromosome domains. The cross-species conserved co-expression clusters identified for ESC-critical pathways suggest fundamental mechanisms of transcription regulation important for ESC development. The promoter analysis identifying conserved transcription factor binding sites on the genes of the conserved co-expression modules provides further evidence of common regulatory mechanism underlying the co-expression. The global co-expression networks provide an overall view of the organization of ESC-critical transcriptomes and help to identify hub genes and transcription factors important in determining the fate of ESC. The chromosomal co-expression illustrates transcriptional events operating on the local chromosome level in ESCs differentiation. The findings and methods presented in this chapter provide novel evidence on the system-level to help understanding how genes interact with each other to regulate the development of ESCs.

CONFLICT OF INTEREST

None declared.

ACKNOWLEDGEMENT

None declared.

DISCLOSURE

Part of information included in this chapter has been previously published in PLoS ONE 3 (10): e3406. doi:10.1371/journal.pone.0003406.

REFERENCES

[1]   Abeyta MJ, Clark AT, Rodriguez RT, *et al.* Unique gene expression signatures of independently-derived human embryonic stem cell lines. Hum Mol Genet 2004; 13 (6):601-8.
[2]   Brandenberger R, Wei H, Zhang S, *et al.* Transcriptome characterization elucidates signaling networks that control human ES cell growth and differentiation. Nat Biotechnol 2004; 22 (6):707-16.
[3]   Liu Y, Shin S, Zeng X, *et al.* Genome wide profiling of human embryonic stem cells (hESCs), their derivatives and embryonal carcinoma cells to develop base profiles of U.S. Federal government approved hESC lines. BMC Dev Biol 2006; 6:20.
[4]   Sato N, Sanjuan IM, Heke M, *et al.* Molecular signature of human embryonic stem cells and its comparison with the mouse. Dev Biol 2003; 260 (2):404-13.
[5]   Skottman H, Mikkola M, Lundin K, *et al.* Gene expression signatures of seven individual human embryonic stem cell lines. Stem Cells 2005; 23 (9):1343-56.
[6]   Li H, Sun Y, Zhan M. The discovery of transcriptional modules by a two-stage matrix decomposition approach. Bioinformatics 2007; 23 (4):473-9.
[7]   Zhan M. Deciphering modular and dynamic behaviors of transcriptional networks. Genomic Med 2007; 1 (1-2):19-28.
[8]   Segal E, Friedman N, Koller D, Regev A. A module map showing conditional activity of expression modules in cancer. Nat Genet 2004; 36 (10):1090-8.
[9]   Barabasi AL, Oltvai ZN. Network biology: understanding the cell's functional organization. Nat Rev Genet 2004; 5 (2):101-13.

[10]   Butte AJ, Tamayo P, Slonim D, Golub TR, Kohane IS. Discovering functional relationships between RNA expression and chemotherapeutic susceptibility using relevance networks. Proc Natl Acad Sci U S A 2000; 97 (22):12182-6.

[11]   Aburatani S, Goto K, Saito S, Toh H, Horimoto K. ASIAN: a web server for inferring a regulatory network framework from gene expression profiles. Nucleic Acids Res 2005; 33 (Web Server issue):W659-64.

[12]   Ihmels J, Bergmann S, Berman J, Barkai N. Comparative gene expression analysis by differential clustering approach: application to the Candida albicans transcription program. PLoS Genet 2005; 1 (3):e39.

[13]   Oldham MC, Horvath S, Geschwind DH. Conservation and evolution of gene coexpression networks in human and chimpanzee brains. Proc Natl Acad Sci U S A 2006; 103 (47):17973-8.

[14]   Sun Y, Li H, Liu Y, *et al.* Evolutionarily Conserved Transcriptional Co-Expression Guiding Embryonic Stem Cell Differentiation. PLoS ONE 2008; 3 (10):e3406.

[15]   Li H, Liu Y, Shin S, *et al.* Transcriptome coexpression map of human embryonic stem cells. BMC Genomics 2006; 7 (1):103.

[16]   Sun Y, Li H, Liu Y, *et al.* Cross-species transcriptional profiles establish a functional portrait of embryonic stem cells. Genomics 2007; 89 (1):22-35.

[17]   Li H, Liu Y, Shin S, *et al.* Transcriptome coexpression map of human embryonic stem cells. BMC Genomics 2006; 7:103.

[18]   Rao M. Conserved and divergent paths that regulate self-renewal in mouse and human embryonic stem cells. Dev Biol 2004; 275 (2):269-86.

[19]   Sun Y, Li H, Yang H, Rao MS, Zhan M. Mechanisms controlling embryonic stem cell self-renewal and differentiation. Crit Rev Eukaryot Gene Expr 2006; 16 (3):211-31.

[20]   Bai Z DJ. Calculating generalized singular value decomposition. SIAM J Sci Comp 1993; 14:1464-86.

[21]   Alter O, Brown PO, Botstein D. Generalized singular value decomposition for comparative analysis of genome-scale expression data sets of two different organisms. Proc Natl Acad Sci U S A 2003; 100 (6):3351-6.

[22]   van der Lann MJ. A new partitioning around medoids algorithm. J Stat Comput Sim 2003; 73:575-84.

[23]   Stead E, White J, Faast R, *et al.* Pluripotent cell division cycles are driven by ectopic Cdk2, cyclin A/E and E2F activities. Oncogene 2002; 21 (54):8320-33.

[24]   Burdon T, Smith A, Savatier P. Signalling, cell cycle and pluripotency in embryonic stem cells. Trends Cell Biol 2002; 12 (9):432-8.

[25]   Fluckiger AC, Marcy G, Marchand M, *et al.* Cell cycle features of primate embryonic stem cells. Stem Cells 2006; 24 (3):547-56.

[26]   Sato N, Meijer L, Skaltsounis L, Greengard P, Brivanlou AH. Maintenance of pluripotency in human and mouse embryonic stem cells through activation of Wnt signaling by a pharmacological GSK-3-specific inhibitor. Nat Med 2004; 10 (1):55-63.

[27]   Miyabayashi T, Teo JL, Yamamoto M, *et al.* Wnt/beta-catenin/CBP signaling maintains long-term murine embryonic stem cell pluripotency. Proc Natl Acad Sci U S A 2007; 104 (13):5668-73.

[28]   Cai L, Ye Z, Zhou BY, *et al.* Promoting human embryonic stem cell renewal or differentiation by modulating Wnt signal and culture conditions. Cell Res 2007; 17 (1):62-72.

[29] Dravid G, Ye Z, Hammond H, *et al.* Defining the role of Wnt/beta-catenin signaling in the survival, proliferation, and self-renewal of human embryonic stem cells. Stem Cells 2005; 23 (10):1489-501.

[30] Humphrey RK, Beattie GM, Lopez AD, *et al.* Maintenance of pluripotency in human embryonic stem cells is STAT3 independent. Stem Cells 2004; 22 (4):522-30.

[31] Yoshida K, Chambers I, Nichols J, *et al.* Maintenance of the pluripotential phenotype of embryonic stem cells through direct activation of gp130 signalling pathways. Mech Dev 1994; 45 (2):163-71.

[32] Paling NR, Wheadon H, Bone HK, Welham MJ. Regulation of embryonic stem cell self-renewal by phosphoinositide 3-kinase-dependent signaling. J Biol Chem 2004; 279 (46):48063-70.

[33] Takahashi K, Murakami M, Yamanaka S. Role of the phosphoinositide 3-kinase pathway in mouse embryonic stem (ES) cells. Biochem Soc Trans 2005; 33 (Pt 6):1522-5.

[34] Ogawa K, Saito A, Matsui H, *et al.* Activin-Nodal signaling is involved in propagation of mouse embryonic stem cells. J Cell Sci 2007; 120 (Pt 1):55-65.

[35] James D, Levine AJ, Besser D, Hemmati-Brivanlou A. TGFbeta/activin/nodal signaling is necessary for the maintenance of pluripotency in human embryonic stem cells. Development 2005; 132 (6):1273-82.

[36] Vallier L, Alexander M, Pedersen RA. Activin/Nodal and FGF pathways cooperate to maintain pluripotency of human embryonic stem cells. J Cell Sci 2005; 118 (Pt 19):4495-509.

[37] Ying QL, Nichols J, Chambers I, Smith A. BMP induction of Id proteins suppresses differentiation and sustains embryonic stem cell self-renewal in collaboration with STAT3. Cell 2003; 115 (3):281-92.

[38] Xu RH, Chen X, Li DS, *et al.* BMP4 initiates human embryonic stem cell differentiation to trophoblast. Nat Biotechnol 2002; 20 (12):1261-4.

[39] Allocco DJ, Kohane IS, Butte AJ. Quantifying the relationship between co-expression, co-regulation and gene function. BMC Bioinformatics 2004; 5:18.

[40] Mitsui K, Tokuzawa Y, Itoh H, *et al.* The homeoprotein Nanog is required for maintenance of pluripotency in mouse epiblast and ES cells. Cell 2003; 113 (5):631-42.

[41] Sharrocks AD. The ETS-domain transcription factor family. Nat Rev Mol Cell Biol 2001; 2 (11):827-37.

[42] van de Wetering M, Cavallo R, Dooijes D, *et al.* Armadillo coactivates transcription driven by the product of the Drosophila segment polarity gene dTCF. Cell 1997; 88 (6):789-99.

[43] Boyer LA, Lee TI, Cole MF, *et al.* Core transcriptional regulatory circuitry in human embryonic stem cells. Cell 2005; 122 (6):947-56.

[44] Loh YH, Wu Q, Chew JL, *et al.* The Oct4 and Nanog transcription network regulates pluripotency in mouse embryonic stem cells. Nat Genet 2006; 38 (4):431-40.

[45] Wasylyk B, Hagman J, Gutierrez-Hartmann A. Ets transcription factors: nuclear effectors of the Ras-MAP-kinase signaling pathway. Trends Biochem Sci 1998; 23 (6):213-6.

[46] Molenaar M, van de Wetering M, Oosterwegel M, *et al.* XTcf-3 transcription factor mediates beta-catenin-induced axis formation in Xenopus embryos. Cell 1996; 86 (3):391-9.

[47] Haremaki T, Tanaka Y, Hongo I, Yuge M, Okamoto H. Integration of multiple signal transducing pathways on Fgf response elements of the Xenopus caudal homologue Xcad3. Development 2003; 130 (20):4907-17.

[48]   Zhan M, Miura T, Xu X, Rao MS. Conservation and variation of gene regulation in embryonic stem cells assessed by comparative genomics. Cell Biochem Biophys 2005; 43 (3):379-405.

[49]   Stuart JM, Segal E, Koller D, Kim SK. A gene-coexpression network for global discovery of conserved genetic modules. Science 2003; 302 (5643):249-55.

[50]   Ravasz E, Somera AL, Mongru DA, Oltvai ZN, Barabasi AL. Hierarchical organization of modularity in metabolic networks. Science 2002; 297 (5586):1551-5.

[51]   van Noort V, Snel B, Huynen MA. The yeast coexpression network has a small-world, scale-free architecture and can be explained by a simple model. EMBO Rep 2004; 5 (3):280-4.

[52]   Jeong H, Mason SP, Barabasi AL, Oltvai ZN. Lethality and centrality in protein networks. Nature 2001; 411 (6833):41-2.

[53]   Qiu XB, Lin YL, Thome KC, *et al.* An eukaryotic RuvB-like protein (RUVBL1) essential for growth. J Biol Chem 1998; 273 (43):27786-93.

[54]   Bauer A, Chauvet S, Huber O, *et al.* Pontin52 and reptin52 function as antagonistic regulators of beta-catenin signalling activity. EMBO J 2000; 19 (22):6121-30.

[55]   Feng Y, Lee N, Fearon ER. TIP49 regulates beta-catenin-mediated neoplastic transformation and T-cell factor target gene induction via effects on chromatin remodeling. Cancer Res 2003; 63 (24):8726-34.

[56]   Wood MA, McMahon SB, Cole MD. An ATPase/helicase complex is an essential cofactor for oncogenic transformation by c-Myc. Mol Cell 2000; 5 (2):321-30.

[57]   Ikura T, Ogryzko VV, Grigoriev M, *et al.* Involvement of the TIP60 histone acetylase complex in DNA repair and apoptosis. Cell 2000; 102 (4):463-73.

[58]   Shen X, Mizuguchi G, Hamiche A, Wu C. A chromatin remodelling complex involved in transcription and DNA processing. Nature 2000; 406 (6795):541-4.

[59]   Alekseev OM, Widgren EE, Richardson RT, O'Rand MG. Association of NASP with HSP90 in mouse spermatogenic cells: stimulation of ATPase activity and transport of linker histones into nuclei. J Biol Chem 2005; 280 (4):2904-11.

[60]   Toyoda M, Kojima M, Takeuchi T. Jumonji is a nuclear protein that participates in the negative regulation of cell growth. Biochem Biophys Res Commun 2000; 274 (2):332-6.

[61]   Jung J, Mysliwiec MR, Lee Y. Roles of JUMONJI in mouse embryonic development. Dev Dyn 2005; 232 (1):21-32.

[62]   Giuliano CJ, Kerley-Hamilton JS, Bee T, *et al.* Retinoic acid represses a cassette of candidate pluripotency chromosome 12p genes during induced loss of human embryonal carcinoma tumorigenicity. Biochim Biophys Acta 2005; 1731 (1):48-56.

[63]   Boyer LA, Plath K, Zeitlinger J, *et al.* Polycomb complexes repress developmental regulators in murine embryonic stem cells. Nature 2006; 441 (7091):349-53.

[64]   Lee TI, Jenner RG, Boyer LA, *et al.* Control of developmental regulators by Polycomb in human embryonic stem cells. Cell 2006; 125 (2):301-13.

[65]   Anneren C, Cowan CA, Melton DA. The Src family of tyrosine kinases is important for embryonic stem cell self-renewal. J Biol Chem 2004; 279 (30):31590-8.

[66]   Meyn MA, 3rd, Schreiner SJ, Dumitrescu TP, Nau GJ, Smithgall TE. SRC family kinase activity is required for murine embryonic stem cell growth and differentiation. Mol Pharmacol 2005; 68 (5):1320-30.

[67]   Fujimoto A, Mitalipov SM, Kuo HC, Wolf DP. Aberrant genomic imprinting in rhesus monkey embryonic stem cells. Stem Cells 2006; 24 (3):595-603.

[68]   Mitalipov S, Clepper L, Sritanaudomchai H, Fujimoto A, Wolf D. Methylation status of imprinting centers for H19/IGF2 and SNURF/SNRPN in primate embryonic stem cells. Stem Cells 2007; 25 (3):581-8.

[69]   Battista S, Pentimalli F, Baldassarre G, *et al.* Loss of Hmga1 gene function affects embryonic stem cell lympho-hematopoietic differentiation. FASEB J 2003; 17 (11):1496-8.

[70]   Martindill DM, Risebro CA, Smart N, *et al.* Nucleolar release of Hand1 acts as a molecular switch to determine cell fate. Nat Cell Biol 2007; 9 (10):1131-41.

[71]   Hughes M, Dobric N, Scott IC, *et al.* The Hand1, Stra13 and Gcm1 transcription factors override FGF signaling to promote terminal differentiation of trophoblast stem cells. Dev Biol 2004; 271 (1):26-37.

[72]   Brimble SN, Zeng X, Weiler DA, *et al.* Karyotypic stability, genotyping, differentiation, feeder-free maintenance, and gene expression sampling in three human embryonic stem cell lines derived prior to August 9, 2001. Stem Cells Dev 2004; 13 (6):585 - 97.

CHAPTER 7

Computational Analysis of Alternative Polyadenylation in Embryonic Stem Cells and Induced Pluripotent Cells

Zhe Ji, Mainul Hoque and Bin Tian*

University of Medicine and Dentistry of New Jersey, USA

Abstract: The 3' untranslated region (3'UTR) of mRNA plays important roles in posttranscriptional control of gene expression. Over half of the human genes have multiple polyadenylation sites in 3'UTRs, leading to 3'UTR isoforms containing different cis elements. Alternative polyadenylation (APA) has been found to be dynamically regulated in different tissue types and under various cellular conditions. Embryonic stem (ES) cells have the ability to self-renew and differentiate into any cell type in the adult body. Posttranscriptional gene regulation through cis elements in 3'UTRs is increasingly found to be important for these functions. In addition, various methods have recently been developed to induce differentiated cells to ES-like cells, called induced pluripotent stem (iPS) cells. Here we show a computational method to examine regulation of 3'UTR by APA using DNA microarray data. We applied this method to ES cells and iPS cells derived from different cell types.

Keywords: mRNA processing, alternative polyadenylation, 3' end formation, embryonic stem cells, induced pluripotent cells, 3'UTR, microRNA, development, proliferation, differentiation, mRNA isoform, post-transcriptional gene regulation.

INTRODUCTION

ES and iPS Cells: Embryonic stem (ES) cells are derived from inner cell mass of the blastocyst, an early stage embryo, and have the property of self-renewal and ability to differentiate into all cell types of the three primary germ layers: ectoderm, endoderm, and mesoderm, corresponding to over 220 types in the adult body. A number of methods have recently been developed to reprogram differentiated somatic cells to ES-like cells using a set of defined factors [1, 2]. These cells, named induced pluripotent stem (iPS) cells, have stimulated the excitement that many ethical and technical barriers associated with clinical

*Address correspondence to B. Tian: Department of Biochemistry and Molecular Biology, Graduate School of Biomedical Sciences and New Jersey Medical School, University of Medicine and Dentistry of New Jersey, Newark, NJ, USA; Tel: (973) 972-3615; Fax: (973) 972-5594; E-mail: btian@umdnj.edu

application of ES cells may be overcome by using iPS cells. In addition, iPS cells have been derived from germ cells under proper culturing conditions, including primordial germ cells from embryos [3] and spermatogonial cells from neonatal or adult testes [4-6].

The 3' Untranslated Regions (3'UTRs): The 3'UTRs of mRNAs contain various cis elements involved in posttranscriptional gene regulation, such as mRNA localization, stability, and translation [7-9]. Some cis elements are present in 3'UTRs of a number of genes, leading to coordinated regulation at the posttranscriptional level, a phenomenon known as RNA regulon [8]. Cis elements with widespread occurrences include AU-rich elements (AREs) [10], GU-rich elements (GREs) [11], and miRNA target sites [12]. They have been shown to play critical roles in ES cell self-renewal and differentiation.

AREs and GREs: AREs, encoded in at least 5% of all genes [10], have widespread role in controlling mRNA stability, including in ES cells [13]. In addition, specific gene regulation *via* AREs have been shown to play roles in different lineages of cell differentiation, including hematopoiesis [14], chondrogenesis [15], myogenesis [16, 17], and neurogenesis [18]. GREs, like AREs, are encoded in a large number of genes [11], and have been suggested to regulate mRNA stability in myogenesis [19].

miRNA Target Sites: miRNAs are ~22 nucleotide (nt) long small RNAs that play important roles in post-transcriptional regulation of gene expression in animals and plants [20]. In animal cells, miRNAs bind to their target sites in mRNAs, mostly in 3'UTRs, which leads to mRNA degradation and/or inhibition of translation [21, 22]. Most mammalian mRNAs have been found to contain conserved miRNA target sites [23]. miRNAs are expressed in a development- and tissue-specific manner and are believed to play roles in sharpening the gene expression program in a particular cell type by inhibiting transcripts not needed in the cell. This property of miRNA has been well documented in ES cells and during their differentiation. Specific sets of miRNAs are expressed in undifferentiated ES cells [24, 25], some of which have been shown to play a role in maintaining the self-renewal capability and inhibition of differentiation [26]. Different sets of miRNAs are expressed when ES cells differentiate into various cell lineages [27], for example miR-1 and miR-133 in myogenesis [28] and miR-124 in neurogenesis [29, 30].

PolyA Sites: The 3' end of almost all mRNAs expressed in eukaryotes is defined by the pre-mRNA polyadenylation site, or polyA site. The 3' end processing machinery, comprising over 90 proteins in human cells [31], are responsible for recognizing the polyA site and carrying out pre-mRNA cleavage and polyadenylation, two tightly coupled reactions also called mRNA polyadenylation [32, 33]. Both upstream and downstream elements surrounding a polyA site are critical for mRNA polyadenylation. For example, the CPSF complex interacts with the upstream A [A/U]UAAA hexamer, also known as the polyadenylation signal (PAS); and the CstF complex interacts with the downstream U-rich and GU-rich elements. In addition, various upstream and downstream auxiliary elements have been found to play regulatory roles in polyA site usage [34, 35].

Regulation of 3'UTRs by Alternative Cleavage and Polyadenylation (APA): Over half of all mammalian genes have multiple polyA sites, resulting in mRNA isoforms with different 3'UTRs and/or coding sequences (CDS) [36, 37]. Compared with constitutive regions of 3'UTRs, alternative regions are usually longer by ~2 fold, have higher AU content, and contain more cis elements [38]. Regulation of APA under various biological conditions has been analyzed for many cellular and viral genes [39, 40]. Recent global analyses have indicated that the APA pattern varies among tissue types [41, 42]. For example, mRNAs expressed in brain tissues tend to have longer 3'UTRs than other tissue types [41], and those expressed in testes tend to have short 3'UTRs and result from polyA sites that are not frequently used in other tissues [41, 43]. In addition, APA can be dynamically regulated in response to extracellular signals, for example activation of neuronal cells [44]. Sandberg and coworkers reported a general trend of 3'UTR shortening in proliferating cells [45]. Mayr and Bartel further found that 3'UTR shortening is more apparent in transformed cells than nontransformed ones with similar proliferating rates [46]. We recently reported that 3'UTRs progressively lengthen *via* APA during mouse embryonic development [38] and this regulation coordinates with various aspects of development including proliferation, differentiation, and morphogenesis.

METHODS

A Computational Method to Analyze 3'UTR Isoforms

To examine how 3'UTRs are regulated by APA, we developed a computational method to analyze expression of 3'UTR isoforms using DNA microarray data. We previously found that a large fraction of probes on Affymetrix GeneChip microarrays are designed to target 3'UTR regions [47]. This design creates an opportunity to examine relative expression of 3'UTR isoforms, because some probes target regions that are regulated by APA, as illustrated in Fig. **1**. For genes with APA, the first and last polyA sites in the 3'-most exon are named proximal and distal sites, respectively, and other sites between them are named middle sites. Accordingly, the regions upstream and downstream of the proximal site are named constitutive UTR (cUTR) and alternative UTR (aUTR), respectively. In addition, the 3'UTR in genes without APA is named single UTR (sUTR). We use polyA sites obtained from PolyA_DB [48] for definition of cUTRs and aUTRs.

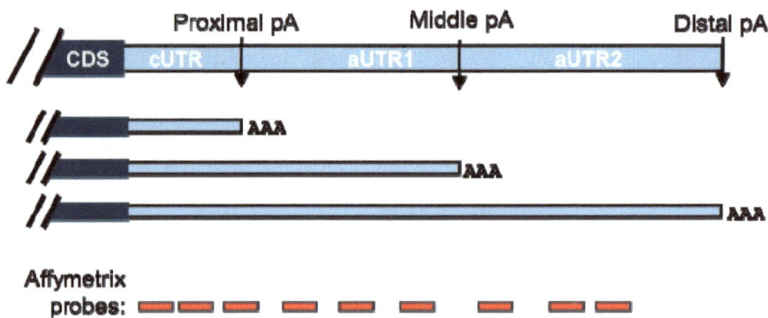

Figure 1. Schematic of APA and analysis of APA using Affymetrix GeneChip probes. (A) A hypothetical gene with 3 polyA sites in the 3'UTR, leading to 3 transcript isoforms. The common region is named constitutive UTR (cUTR) and the alternative region alternative UTR (aUTR). The first and last polyA sites are named proximal and distal polyA sites and the one in the middle is named middle polyA site. The middle polyA site divides aUTR into aUTR1 and aUTR2. Affymetrix GeneChip probes targeting cUTRs and aUTRs are shown as red bars. CDS, coding sequence; AAA, poly (A) tail.

Our method to analyze 3'UTR regulation is based on comparison of microarray probes targeting different regions in 3'UTRs. Intuitively, higher probe intensities for the aUTR indicates more expression of long 3'UTR isoforms. However, because of different hybridization properties, probes in different regions cannot be directly compared at the intensity level. We thus obtain ratio for each probe based on comparison with the median of all samples. Probe ratios are averaged for probes corresponding to a given UTR subregion, *i.e.,* cUTR or aUTR. The averaged ratio of cUTR is then compared with that of aUTR to infer relative change of expression of 3'UTR isoforms (Fig. **2**). We map Affymetrix GeneChip probes to cUTR and aUTR sequences using BLAST.

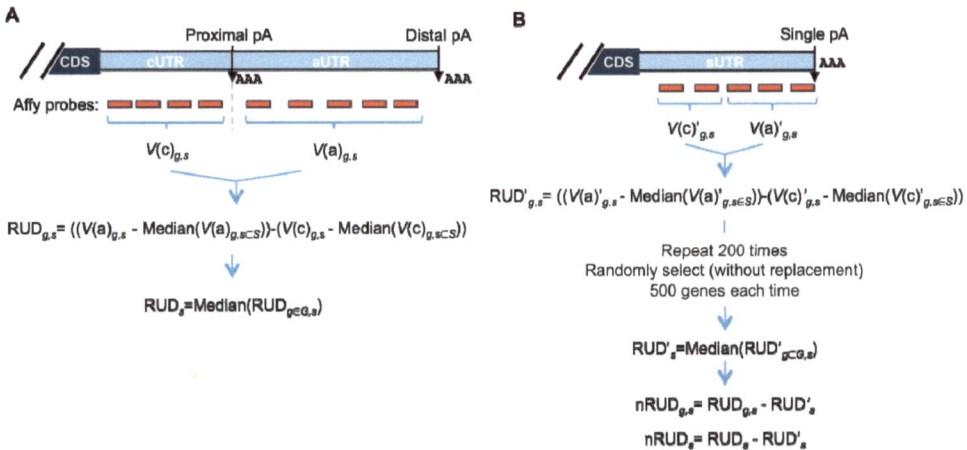

Figure 2: Calculation and normalization of RUD. (A) Calculation of RUD. cUTR, constitutive UTR; aUTR, alternative UTR; CDS, coding sequence. **(B)** Calculation of RUD' and nRUD using genes without APA. sUTR, single UTR; AAA, poly (A) tail. Red bars are Affymetrix GeneChip probes. See text for detailed description of the methods.

The ratio of probe ratio of the downstream region of a polyA site to that of the upstream region is called Relative Usage of Distal polyA site, or RUD. This score reflects relative usage of a polyA site. A large RUD value indicates high expression of 3'UTR isoforms resulting from downstream polyA site (s), whereas a small RUD indicates low expression of these isoforms. RUD can be calculated for each proximal or middle polyA site that is amenable to such analysis. In addition, the median RUD of all surveyed polyA sites can be used to represent the global trend of polyA site usage and 3'UTR length. The detailed calculation of RUD is as follows (also illustrated in Fig. **2**):

Let's say there is a sample set S containing s number of samples belonging to different biological groups. Each sample is denoted by s. G is an analyzed gene set containing g number of genes. Each gene is denoted by g. V is $\log_2$ (probe intensity). $V(c_1)_{g,s}$, $V(c_2)_{g,s}$, ... $V(c_n)_{g,s}$ are V for probes 1, 2, ...n in the cUTR of gene g in sample s. $V(a_1)_{g,s}$, $V(a_2)_{g,s}$, ... $V(a_n)_{g,s}$ are V for probes 1, 2, ...n in the aUTR of gene g in sample s. We first calculate $V(c)_{g,s}$ and $V(a)_{g,s}$, which are averaged $\log_2$ (probe intensity) for cUTR and aUTR, respectively. Thus, $V(c)_{g,s} =$ mean $(V(c_1)_{g,s}, V(c_2)_{g,s}, ... V(c_n)_{g,s})$ and $V(a)_{g,s} =$ mean $(V(a_1)_{g,s}, V(a_2)_{g,s}, ... V(a_n)_{g,s})$. Next, $V(c)_{g,s}$ and $V(a)_{g,s}$ of each sample are normalized to their respective median across all samples, *i.e.,* Median $(V(c)_{g,s\in S})$ and Median $(V(a)_{g,s\in S})$. RUD for each gene in a sample is calculated by $RUD_{g,s} = (V(a)_{g,s} -$ Median $(V(a)_{g,s\in S})) - (V(c)_{g,s} -$ Median $(V(c)_{g,s\in S}))$. And RUD for each sample is calculated by $RUD_s =$ Median $(RUD_{g\in G,s})$.

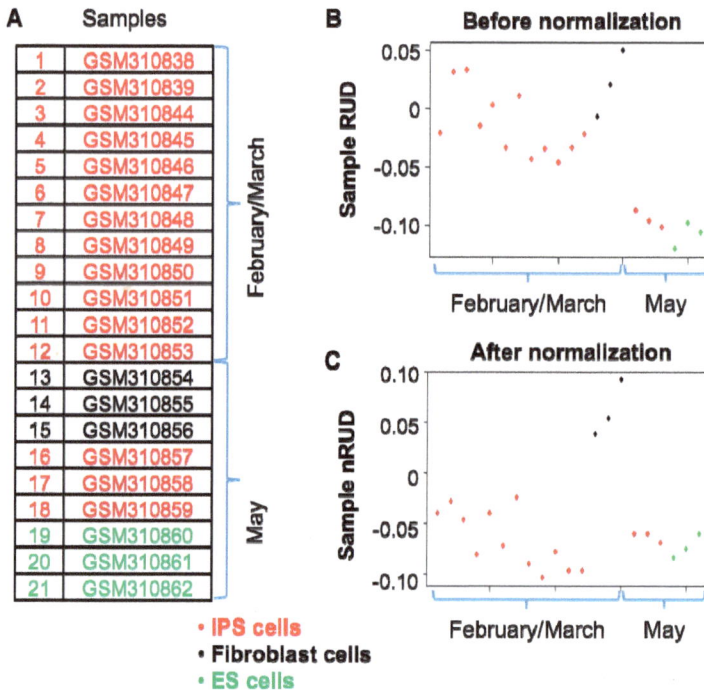

Figure 3: Correction of systematic differences between data sets using the nRUD method. (A) Data set for the generation of iPS cells from human BJ fibroblast. Sample names are shown as their GEO Accession numbers and are colored according to their biological groups, *i.e.,* iPS cells (red), fibroblast cells (black), and ES cells (green). Samples were processed at different times, including the Feburary/March batch and May batch. **(B)** RUD values without normalization. Samples are in the same order as (A). **(C)** nRUD values, *i.e.,* RUD values after normalization. Samples are in the same order as (A). Samples in the same biological group have closer nRUD values than original RUD values.

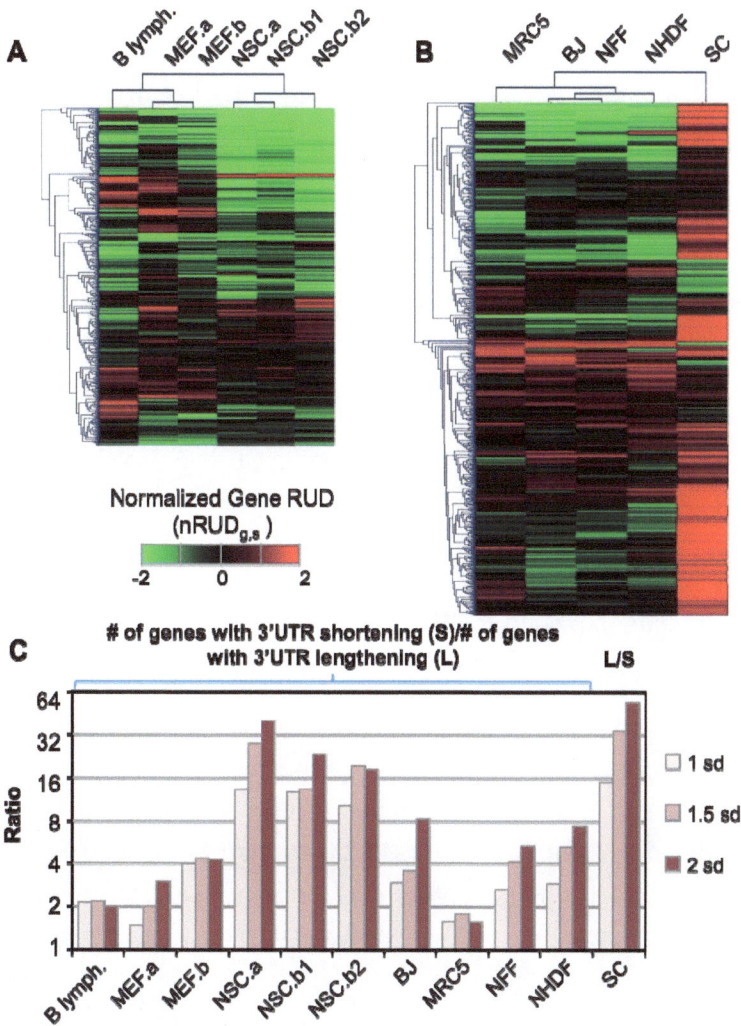

Figure 4: Regulation of 3'UTR by APA in generation of iPS cells across different sample sets. (A) Heatmap showing 674 mouse genes with APA surveyed in 6 sample sets. The 3'UTR regulation was measured by gene nRUD, which is represented by color according to the scale shown in the graph, with red indicating 3'UTR lengthening and green 3'UTR shortening. Samples and genes are also clustered using gene nRUD by hierarchical clustering and Pearson Correlation. B lymph. is for B lymphocyte; MEF.a and MEF.b are two separate data sets for mouse embryonic firoblasts; NSC.a, NSC.b1 and NSC.b2 are three separate data sets for adult neural stem cells (NSCs). **(B)** As in (A), 996 human genes were surveyed. BJ, NFF, and NHDF are for neonatal foreskin fibroblast cells; MRC5 is for fetal lung fibroblast cells; SC is for spermatogonial cells. **(C)** Ratio of number of genes with 3'UTR shortening (S) to number of genes with 3'UTR lengthening (L) in each condition. Selection of genes with 3'UTR regulation in generation of iPS cells is based on 3 cutoffs, *i.e.,* 1x, 1.5x, and 2x standard deviation of all nRUD values for each condition. L/S is used for SC.

RUD Normalization

One important observation we have made is that different batches of microarray data could have systematic differences in RUD (Fig. **3**). This is likely to be due to technical variations in sample processing. For example, different protocols of reverse transcription can result in cDNAs of different sizes. Those reactions leading to short cDNAs can potentially cause overestimation of long 3'UTR isoforms because cUTR sequences are relatively less reverse transcribed. Similarly, different protocols of *in vitro* transcription and aRNA fragmentation can differentially affect hybridization for probes in cUTRs and aUTRs.

To address systematic variations between samples, we have developed a method to normalize RUD values across samples by using probes targeting sUTRs (Fig. **2B**). Our rationale is that comparison of 5' probes with 3' ones in sUTR can provide a background difference for probes in cUTRs and aUTRs, controlling for RUD variations between samples. We randomly select two 5' and two 3' probes in sUTRs and calculate expected RUD, named RUD'. We repeat this process 200 times, and select 500 genes each time. The detailed calculation is as follows:

Let's denote all the values derived from randomly sampled probes as expected values. $V(c)'$ is expected $V(c)$ and $V(a)'$ is expected $V(a)$. They are calculated by $V(c)'_{g,s} = (V(c_1)'_{g,s} + V(c_2)_{g,s}')/2$ and $V(a)'_{g,s} = (V(a_1)'_{g,s} + V(a_2)'_{g,s})/2$. $RUD'_{g,s}$ is calculated by $RUD'_{g,s} = (V(a)'_{g,s} - \text{Median}(V(a)'_{g,s \in S})) - (V(c)'_{g,s} - \text{Median}(V(c)'_{g,s \in S}))$. Note the gene g here is randomly selected from genes with sUTRs. RUD'_s is calculated by $RUD'_s = \text{Median}(RUD'_{g \in G,s})$.

The normalized RUD, or nRUD, is calculated by $nRUD_{g,s} = RUD_{g,s} - RUD'_s$ and $nRUD_s = RUD_s - RUD'_s$. As shown in Fig. **3**, this approach significantly reduces sample-to-sample variations in RUD calculation: biologically similar samples processed at different times have closer RUD values after normalization than before. In addition, the 3'UTR length in reprogrammed iPS cells is closer to ES cells than to source fibroblast cells, consistent with their phenotypes and gene expression profiles (Fig. **3**).

Analysis of 3'UTR Regulation in Generation of iPS Cells

We applied our method to a set of DNA microarray data for ES cells and iPS cells reprogrammed from different somatic cells. We included 5 data sets for mouse iPS cells derived from B lymphocytes (B lymph.) [49], mouse embryonic fibroblasts (MEFs) [50], and adult neural stem cells (NSCs) [51, 52], and 4 data sets of human iPS cells derived from neonatal foreskin fibroblasts (BJ, NFF, and NHDF) and fetal lung fibroblasts (MRC5) [53-56]. In addition, we included a data set for human iPS cells derived from spermatogonial cells (SC), a type of germ cell from human adult testis [6].

Using nRUD we found that in general genes tend to express mRNAs with shorter 3'UTRs in iPS cells than in source somatic cells (Fig. **4**). By contrast, 3'UTRs lengthen during generation of iPS cells from SC. In addition, the extent of 3'UTR shortening appears to be variable for different cell types (Fig. **4C**). Reprogramming of NSC involves more drastic 3'UTR shortening than other cell types, which is similar in extent, but opposite in direction, to reprogramming of SC (Fig. **4C**). This result indicates that the direction and extent of 3'UTR regulation in generation of iPS cells reflect the difference between source cells and iPS cells. In addition, each cell type has a set of genes with a different direction of 3'UTR regulation than the global trend of the cell (Fig. **4A** and **4B**), suggesting cell-specific regulation of APA for certain genes.

DISCUSSION

We show that 3'UTRs are reprogrammed by APA during generation of iPS cells from different cell types. This trend is opposite to the regulation in embryonic development, during which 3'UTRs progressively lengthen [38]. In addition, lengthening of 3'UTRs in reprogramming of germ cells appears to be opposite to the trend in postnatal development of testis [38]. These results underline the dynamic nature of 3'UTR regulation in development, and indicate that APA is an integral part of cell reprogramming process. Since longer 3'UTRs are more likely to be regulated at the post-transcriptional level, APA can play an important role in proliferation and differentiation of ES cells. Consistently, we found that long 3'UTR isoforms containing target sites for ES cell-specific miRNAs are more likely to be inhibited than short isoforms without the target sites [57]. It remains to be seen whether the 3'UTR length can be used as a marker to monitor cell reprogramming process, and whether perturbation of APA may alter the efficiency of generation of iPS cells.

Our RUD values are based on microarray probes targeting cUTR and aUTRs that are defined by proximal and distal polyA sites. About 40-50% human and mouse genes with APA contain also polyA sites between proximal and distal sites, *i.e.,* middle polyA sites. Analysis of middle polyA sites would require dividing aUTR probes into different groups according to middle polyA site positions. For example, if there is a middle polyA site, the region between the proximal polyA site and middle polyA site will be aUTR1 and the region between the distal polyA site and middle polyA site will be aUTR2 (Fig. **1**). To analyze 3'UTR isoforms resulting from the middle polyA site, an RUD value can be calculated by comparing probes corresponding to aUTR1 and aUTR2. On the other hand, recent advances in the deep sequencing technology has enabled analysis of mRNA isoforms at a single nucleotide resolution [42, 58]. We expect APA analysis will greatly benefit from deep sequencing approaches in the future.

We show different somatic cell types have different extent of 3'UTR regulation despite a consistent overall trend. Whereas noise in analyzing heterogeneous data sets can be a contributing factor, cell-specific regulation is very likely to be the main reason for variation in 3'UTR regulation. In support of this notion, a growing number of factors have been shown to regulate polyadenylation [39]. For example, the RNA binding protein Nova has been shown to have a widespread impact on APA in neuronal cells [59]. Conversely, since the APA pattern is determined by both cell type and developmental state, it can be used as biomarker for sample clustering and classification. Since the APA pattern is calculated by comparing different isoforms, the data are internally normalized, and can be more robust than mRNA levels for separating samples.

CONFLICT OF INTEREST

None declared.

ACKNOWLEDGEMENTS

We thank members of BT lab for helpful discussions. This work was funded by an NIH grant (R01 GM084089).

DISCLOSURE

Part of information included in this chapter has been previously published in PLoS ONE 4 (12): e8419.

REFERENCES

1]   Takahashi K, Yamanaka S. Induction of pluripotent stem cells from mouse embryonic and adult fibroblast cultures by defined factors. Cell 2006; 126 (4):663-76.

[2]   Yu J, Thomson JA. Pluripotent stem cell lines. Genes Develop 2008; 22 (15):1987-97.

[3]   Matsui Y, Zsebo K, Hogan BL. Derivation of pluripotential embryonic stem cells from murine primordial germ cells in culture. Cell 1992; 70 (5):841-7.

[4]   Kanatsu-Shinohara M, Inoue K, Lee J, *et al.* Generation of pluripotent stem cells from neonatal mouse testis. Cell 2004; 119 (7):1001-12.

[5]   Guan K, Nayernia K, Maier LS, *et al.* Pluripotency of spermatogonial stem cells from adult mouse testis. Nature 2006; 440 (7088):1199-203.

[6]   Conrad S, Renninger M, Hennenlotter J, *et al.* Generation of pluripotent stem cells from adult human testis. Nature 2008; 456 (7220):344-9.

[7]   Wickens M, Anderson P, Jackson RJ. Life and death in the cytoplasm: messages from the 3' end. Curr Opin Genet Dev 1997; 7 (2):220-32.

[8]   Keene JD. RNA regulons: coordination of post-transcriptional events. Nat Rev Genet 2007; 8 (7):533-43.

[9]   Garneau NL, Wilusz J, Wilusz CJ. The highways and byways of mRNA decay. Nat Rev Mol Cell Biol 2007; 8 (2):113-26.

[10]   Bakheet T, Williams BR, Khabar KS. ARED 3.0: the large and diverse AU-rich transcriptome. Nucleic Acids Research 2006; 34 (Database issue):D111-4.

[11]   Vlasova IA, Tahoe NM, Fan D, *et al.* Conserved GU-rich elements mediate mRNA decay by binding to CUG-binding protein 1. Mol Cell 2008; 29 (2):263-70.

[12]   Lewis BP, Burge CB, Bartel DP. Conserved seed pairing, often flanked by adenosines, indicates that thousands of human genes are microRNA targets. Cell 2005; 120 (1):15-20.

[13]   Sharova LV, Sharov AA, Nedorezov T, *et al.* Database for mRNA half-life of 19 977 genes obtained by DNA microarray analysis of pluripotent and differentiating mouse embryonic stem cells. DNA Res 2009; 16 (1):45-58.

[14]   Stumpo DJ, Broxmeyer HE, Ward T, *et al.* Targeted disruption of Zfp36l2, encoding a CCCH tandem zinc finger RNA-binding protein, results in defective hematopoiesis. Blood 2009; 114 (12):2401-10.

[15]   Mukudai Y, Kubota S, Kawaki H, *et al.* Posttranscriptional regulation of chicken ccn2 gene expression by nucleophosmin/B23 during chondrocyte differentiation. Mol Cell Biol 2008; 28 (19):6134-47.

[16]   Li XL, Andersen JB, Ezelle HJ, Wilson GM, Hassel BA. Post-transcriptional regulation of RNase-L expression is mediated by the 3'-untranslated region of its mRNA. J Biol Chem 2007; 282 (11):7950-60.

[17]   Deschenes-Furry J, Belanger G, Mwanjewe J, *et al.* The RNA-binding protein HuR binds to acetylcholinesterase transcripts and regulates their expression in differentiating skeletal muscle cells. J Biol Chem 2005; 280 (27):25361-8.

[18] Ratti A, Fallini C, Cova L, *et al.* A role for the ELAV RNA-binding proteins in neural stem cells: stabilization of Msi1 mRNA. J Cell Sci 2006; 119 (Pt 7):1442-52.

[19] Lee JE, Lee JY, Wilusz J, Tian B, Wilusz CJ. Systematic analysis of cis-elements in unstable mRNAs demonstrates that CUGBP1 is a key regulator of mRNA decay in muscle cells. PLoS One 2010; 5 (6):e11201.

[20] Bartel DP. MicroRNAs: genomics, biogenesis, mechanism, and function. Cell 2004; 116 (2):281-97.

[21] Bartel DP. MicroRNAs: target recognition and regulatory functions. Cell 2009; 136 (2):215-33.

[22] Filipowicz W, Bhattacharyya SN, Sonenberg N. Mechanisms of post-transcriptional regulation by microRNAs: are the answers in sight? Nat Rev Genet 2008; 9 (2):102-14.

[23] Friedman RC, Farh KK, Burge CB, Bartel DP. Most mammalian mRNAs are conserved targets of microRNAs. Genome Res 2009; 19 (1):92-105.

[24] Houbaviy HB, Murray MF, Sharp PA. Embryonic stem cell-specific MicroRNAs. Dev Cell 2003; 5 (2):351-8.

[25] Suh MR, Lee Y, Kim JY, *et al.* Human embryonic stem cells express a unique set of microRNAs. Dev Biol 2004; 270 (2):488-98.

[26] Melton C, Judson RL, Blelloch R. Opposing microRNA families regulate self-renewal in mouse embryonic stem cells. Nature 2010; 463 (7281):621-6.

[27] Ivey KN, Muth A, Arnold J, *et al.* MicroRNA regulation of cell lineages in mouse and human embryonic stem cells. Cell Stem Cell 2008; 2 (3):219-29.

[28] Zhao Y, Ransom JF, Li A, *et al.* Dysregulation of cardiogenesis, cardiac conduction, and cell cycle in mice lacking miRNA-1-2. Cell 2007; 129 (2):303-17.

[29] Cao X, Pfaff SL, Gage FH. A functional study of miR-124 in the developing neural tube. Genes Dev 2007; 21 (5):531-6.

[30] Makeyev EV, Zhang J, Carrasco MA, Maniatis T. The MicroRNA miR-124 promotes neuronal differentiation by triggering brain-specific alternative pre-mRNA splicing. Mol Cell 2007; 27 (3):435-48.

[31] Shi Y, Di Giammartino DC, Taylor D, *et al.* Molecular architecture of the human pre-mRNA 3' processing complex. Mol Cell 2009; 33 (3):365-76.

[32] Colgan DF, Manley JL. Mechanism and regulation of mRNA polyadenylation. Genes Dev 1997; 11 (21):2755-66.

[33] Zhao J, Hyman L, Moore C. Formation of mRNA 3' ends in eukaryotes: mechanism, regulation, and interrelationships with other steps in mRNA synthesis. Microbiol Mol Biol Rev 1999; 63 (2):405-45.

[34] Hu J, Lutz CS, Wilusz J, Tian B. Bioinformatic identification of candidate cis-regulatory elements involved in human mRNA polyadenylation. RNA 2005; 11 (10):1485-93.

[35] Danckwardt S, Hentze MW, Kulozik AE. 3' end mRNA processing: molecular mechanisms and implications for health and disease. Embo J 2008; 27 (3):482-98.

[36] Tian B, Hu J, Zhang H, Lutz CS. A large-scale analysis of mRNA polyadenylation of human and mouse genes. Nucleic Acids Res 2005; 33 (1):201-12.

[37] Yan J, Marr TG. Computational analysis of 3'-ends of ESTs shows four classes of alternative polyadenylation in human, mouse, and rat. Genome Res 2005; 15 (3):369-75.

[38] Ji Z, Lee JY, Pan Z, Jiang B, Tian B. Progressive lengthening of 3' untranslated regions of mRNAs by alternative polyadenylation during mouse embryonic development. Proc Natl Acad Sci U S A 2009; 106:7028-33.

[39] Lutz CS. Alternative polyadenylation: a twist on mRNA 3' end formation. ACS Chem Biol 2008; 3 (10):609-17.

[40] Edwalds-Gilbert G, Veraldi KL, Milcarek C. Alternative poly (A) site selection in complex transcription units: means to an end? Nucleic Acids Res 1997; 25 (13):2547-61.

[41] Zhang H, Lee JY, Tian B. Biased alternative polyadenylation in human tissues. Genome Biol 2005; 6:R100.

[42] Wang ET, Sandberg R, Luo S, *et al.* Alternative isoform regulation in human tissue transcriptomes. Nature 2008; 456 (7221):470-6.

[43] Liu D, Brockman JM, Dass B, *et al.* Systematic variation in mRNA 3'-processing signals during mouse spermatogenesis. Nucleic Acids Res 2007; 35 (1):234-46.

[44] Flavell SW, Kim TK, Gray JM, *et al.* Genome-wide analysis of MEF2 transcriptional program reveals synaptic target genes and neuronal activity-dependent polyadenylation site selection. Neuron 2008; 60 (6):1022-38.

[45] Sandberg R, Neilson JR, Sarma A, Sharp PA, Burge CB. Proliferating cells express mRNAs with shortened 3' untranslated regions and fewer microRNA target sites. Science 2008; 320 (5883):1643-7.

[46] Mayr C, Bartel DP. Widespread Shortening of 3'UTRs by Alternative Cleavage and Polyadenylation Activates Oncogenes in Cancer Cells. Cell 2009; 138 (4):673-84.

[47] D'Mello V, Lee JY, MacDonald CC, Tian B. Alternative mRNA polyadenylation can potentially affect detection of gene expression by affymetrix genechip arrays. Appl Bioinformatics 2006; 5 (4):249-53.

[48] Lee JY, Yeh I, Park JY, Tian B. PolyA_DB 2: mRNA polyadenylation sites in vertebrate genes. Nucleic Acids Res 2007; 35 (Database issue):D165-8.

[49] Mikkelsen TS, Hanna J, Zhang X, *et al.* Dissecting direct reprogramming through integrative genomic analysis. Nature 2008; 454 (7200):49-55.

[50] Sridharan R, Tchieu J, Mason MJ, *et al.* Role of the murine reprogramming factors in the induction of pluripotency. Cell 2009; 136 (2):364-77.

[51] Kim JB, Zaehres H, Wu G, *et al.* Pluripotent stem cells induced from adult neural stem cells by reprogramming with two factors. Nature 2008; 454 (7204):646-50.

[52] Kim JB, Sebastiano V, Wu G, *et al.* Oct4-induced pluripotency in adult neural stem cells. Cell 2009; 136 (3):411-9.

[53] Maherali N, Ahfeldt T, Rigamonti A, *et al.* A high-efficiency system for the generation and study of human induced pluripotent stem cells. Cell Stem Cell 2008; 3 (3):340-5.

[54] Park IH, Zhao R, West JA, *et al.* Reprogramming of human somatic cells to pluripotency with defined factors. Nature 2008; 451 (7175):141-6.

[55] Masaki H, Ishikawa T, Takahashi S, *et al.* Heterogeneity of pluripotent marker gene expression in colonies generated in human iPS cell induction culture. Stem Cell Res 2007; 1 (2):105-15.

[56] Lowry WE, Richter L, Yachechko R, *et al.* Generation of human induced pluripotent stem cells from dermal fibroblasts. Proc Natl Acad Sci U S A 2008; 105 (8):2883-8.

[57] Ji Z, Tian B. Reprogramming of 3' untranslated regions of mRNAs by alternative polyadenylation in generation of pluripotent stem cells from different cell types. PLoS One 2009; 4 (12):e8419.

[58] Pan Q, Shai O, Lee LJ, Frey BJ, Blencowe BJ. Deep surveying of alternative splicing complexity in the human transcriptome by high-throughput sequencing. Nat Genet 2008; 40 (12):1413-5.

[59]   Licatalosi DD, Mele A, Fak JJ, *et al.* HITS-CLIP yields genome-wide insights into brain alternative RNA processing. Nature 2008; 456 (7221):464-9.

CHAPTER 8

Genomics of Alternative Splicing in Stem Cells

Stephanie C. Huelga and Gene W. Yeo[*]

University of California, San Diego, USA

Abstract: Alternative splicing has the ability to expand proteome diversity among different cell and tissue types, and various stages of differentiation and development. In higher eukaryotes, alternative pre-mRNA splicing is highly regulated by many cellular factors and only in the past decade has the magnitude of this co/post-transcriptional regulation been revealed. Here we review recent technologies that have enabled genome-wide detection of alternative splicing, and highlight computational methods that have been developed to identify and understand the interplay between the *cis* and *trans* factors important for regulating alternative splicing in stem cells.

Keywords: Alternative splicing, spliceosome, exons, introns, pre-mRNA, splicing factors, splicing regulatory elements (SRE), RNA binding proteins (RBPs), CLIP-seq or HITS-CLIP, RNA-seq, microarrays, transcriptome, genomics, high-throughput sequencing, splicing code.

INTRODUCTION

Gene expression regulation has been shown in recent years to be far more complex than previously anticipated. Human gene expression is finely regulated for cell-, tissue-, and condition-specific functions *via* a variety of co- and post-transcriptional processes. Alternative splicing is a major mechanism by which the complexity of the proteome is enhanced through the selective excision of non-coding intronic regions of a premature messenger RNA (pre-mRNA) and juxtaposition of exonic regions in different arrangements. From a single pre-mRNA transcript, alternative splicing can generate an array of mRNA isoforms (Fig. **1**).

The spliceosome is a large RNA-protein complex that is essential for pre-mRNA splicing. This complex is made up of five small nuclear ribonucleoprotein particles (snRNPs), U1, U2, U4, U5, and U6, as well as auxiliary factors that

*Address correspondence to G.W. Yeo:** Cellular and Molecular Medicine, University of California San Diego, 2880 Torrey Pines Scenic Drive, Rm 3805, La Jolla, CA 92037, USA; Tel: (858) 534-9321; Fax: (858) 246-1579; E-mail: geneyeo@ucsd.edu

Ming Zhan (Ed)

cooperate to recognize splice sites and catalyze the splicing reaction (Fig. **2**). Interactions between the spliceosomal machinery, *cis*-regulatory elements and *trans*-regulatory splicing factors define a set of rules for accurate splice site selection. While this "splicing code" is still being deciphered, some important splicing regulatory elements (SREs) have been identified. Exonic and intronic splicing enhancer (ESE, ISE) and splicing silencer (ESS, ISS) motifs act to enhance or inhibit adjacent splice site recognition. Trans-acting splicing factors such as the heterogeneous nuclear ribonucleoproteins (hnRNPs), serine-arginine (SR) proteins, and other RNA binding proteins (RBPs) are recruited to these elements to enhance or repress splicing [1-3]. As these splicing factors are differentially expressed in various cell types, their combinatorial interaction regulates the expressed, cell-type specific alternative splicing patterns. Differential expression of these splicing factors, amongst others, allow for alternative splicing to differentially occur in various cell states.

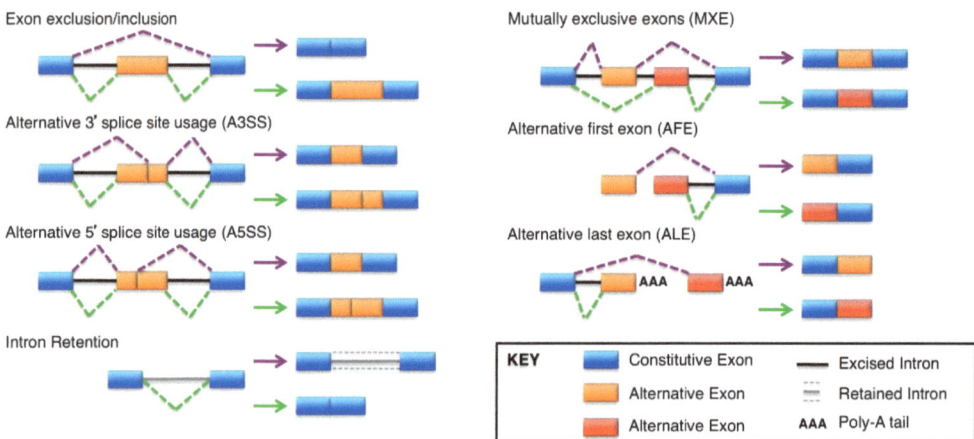

Figure 1: Alternative splicing produces multiple mRNA isoforms from a single pre-mRNA transcript. The events depicted include: exon exclusion/inclusion, alternative 3' splice site usage (A3SS), alternative 5' splice site (A5SS), intron retention, mutually exclusive exons (MXE), alternative first exon (AFE), and alternative last exon (ALE).

It has been suggested that almost all multi-exon human genes undergo alternative splicing [4, 5]. Unexpectedly, genes important for the biology of stem cells have not escaped regulation by alternative splicing. As an example, *OCT4*, which when over expressed in differentiated cells aids in reprogramming cells to a pluripotent state [6], is regulated by alternative splicing. Differential expression of various *OCT4* isoforms has been observed before and after differentiation, each having unique functions, suggesting regulatory roles for stem cell pluripotency [7-9]. Other genes implicated in stem cell biology with functionally distinct isoforms produced by alternative splicing include: insulin-like growth factor 1 (*IGF-1*) [10], DNA methyltransferase 3 Beta (*DNMT3B*) [11], protein kinase C delta (*PKCδ*) [12], and vascular endothelial growth factor A (*VEGFA*) [13].

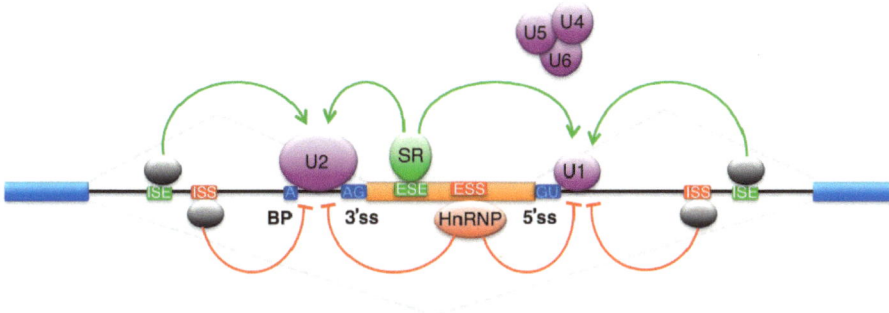

Figure 2: The spliceosome complex and regulatory splicing factors surrounding an alternatively spliced exon. Spliceosomal assembly is initiated by the recognition of the 5' GU by the U1 snRNP. The U2 snRNP is then recruited to the branch-point (BP) in an ATP-dependent manner. Subsequent recruitment of the U4/U6-U5 tri-snRNP is then followed by a series of structural rearrangements leading to the formation of a catalytically active spliceosome complex to remove the alternative exon (orange) and introns (black). Other auxiliary factors (gray) bind to *cis*-regulatory elements, such as SR proteins (green) and the hnRNP proteins (red) that bind to ESE and ESS to enhance or repress splicing, respectively.

Identifying alternative splicing events in stem cells and understanding how they are regulated has far-reaching implications to our understanding of human development and disease. In recent years genome-wide approaches and computational approaches have been created to study alternative splicing and the various *cis*- and *trans*-factors that regulate this process.

GENOME-WIDE METHODS FOR ANALYSIS OF ALTERNATIVE SPLICING AND REGULATORY SPLICING FACTORS

Recent genome-wide approaches have begun to reveal the full extent of alternative splicing in human stem cell differentiation [14, 15]. Prior to these global methods, discovering new alternative splicing events was limited to candidate genes. The availability of microarray and sequencing technologies have remarkably increased our throughput of identifying new alternative splicing events and providing systematic readouts of their biological implication.

Microarrays specifically designed for studying alternative splicing contain exonic oligonucleotide probes and probes spanning two annotated exons that may be spliced together in a specific transcript variant [16]. These "exon-junction arrays" can then be probed with cDNA (complimentary DNA) to measure the degree of alternative splicing (Fig. **3A**). This platform is a useful tool for uncovering tissue-specific splicing patterns [17], and has also revealed many RBP-dependent splicing events *via* comparison of cDNA from wildtype and splicing factor depleted cells [18].

While useful for studying splicing at known exon coordinates and junctions, exon-junction arrays are limited in their ability to detect novel splice sites and are thus biased in alternative splicing detection. Whole genome tiling arrays with overlapping or non-overlapping oligonucleotide probes that cover entire genomic regions of interest circumvent these issues (Fig. **3A**). Unlike exon-junction arrays, these arrays can detect altered splice site usage, retained introns, and unannotated exons [19]. A cost-effective alternative to genome tiling arrays, which exclude intronic and intergenic regions, are exon-tiling arrays [20] (Fig. **3A**). We were among the first to apply these all-exon arrays to explore alternative splicing important in stem cell biology [14, 15]. Our studies revealed that alternative splicing events important in neuronal differentiation of human embryonic stem cells (hESCs) occurs primarily in groups of genes such as serine/threonine kinases and helicases [14]. Another study utilized the same platform to analyze undifferentiated hESCs and derived cardiac progenitor cells [15], revealing splicing events specific to cardiac progenitor cells in gene groups such as cytoskeleton remodeling, RNA splicing, muscle specification, cell cycle checkpoint control, and serine/threonine kinases and helicases.

Profiling methods that rely on microarrays have been developed to identify transcripts bound by specific splicing factors. Ribonucleoprotein (RNP) immunoprecipitation followed by microarray application (RIP-Chip) uses an antibody for a regulatory RBP of interest to pull down an endogenous ribonucleoprotein complex from cell extracts. The bound RNA is extracted from the eluate, purified, and identified by microarray analysis [21]. This method has been successful in revealing sets of mRNA targets that are regulated by similar factors.

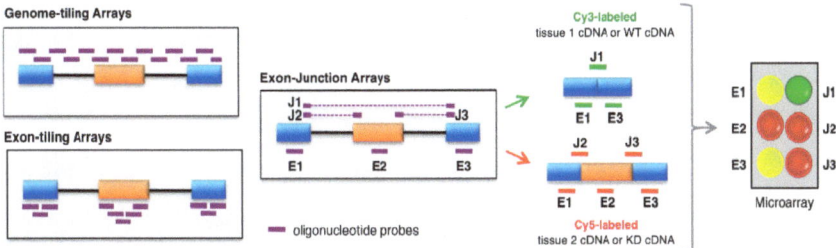

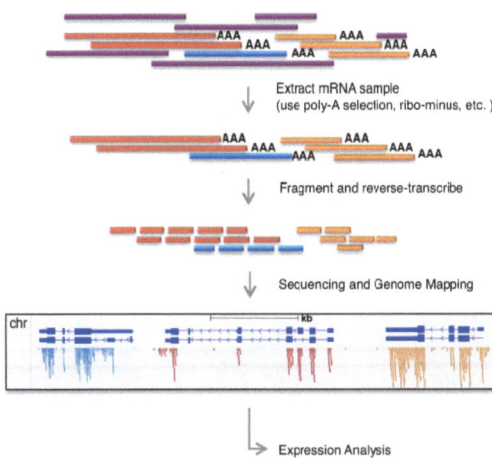

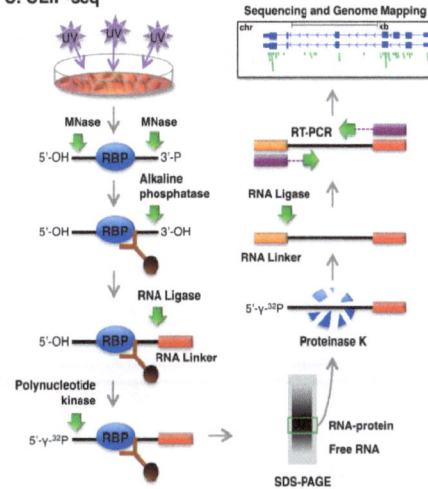

Figure 3: A summary of genome-wide approaches for alternative splicing analysis. A) Microarrays with various oligonucleotide probe types have been applied to study alternative splicing. Whole-genome tiling arrays contain overlapping or non-overlapping probes that scan entire genomic regions. Exon-tiling, or all-exon arrays, contain overlapping or non-overlapping probes within each known or predicted exon of the genome. These tiling arrays are often assayed with biotin-labeled cDNA and then incubated with fluorescently labeled streptavidin for detection. Exon-junction microarrays contain probes for exons (E1, E2, E3) and putative exon-junctions (J1, J2, J3) within a transcript. Cy3 (green) and Cy5 (red) labeling of cDNA from two different samples allows for comparison of expressed transcript variants. Green or red spots on the microarray represent expression exclusively in the Cy3-labeled or Cy5-labeled samples, respectively, while yellow spots on the microarray represent equivalent expression between samples. **B)** The RNA-seq protocol begins with total RNA extraction from a cellular condition or tissue type. The RNA sample is enriched for mRNA with poly-A selection using poly-T oligos and ribo-minus selection to remove contaminating ribosomal RNA, and other selection methods. The

mRNA is then fragmented and reverse transcribed. Lastly, the library is amplified, sequenced and mapped to the genome. **C)** The CLIP-seq protocol begins with UV-crosslinking of RBPs to bound RNA targets and the RBP-RNA complex is enriched using a specific antibody. Trimming with MNase enables nucleotide level resolution as the RNA bound by the protein is protected from digestion. After size selection, the protein is removed from its bound RNA and the RNA is amplified by RT-PCR, sequenced and mapped to the genome to determine significant regions of protein binding for the RBP.

While powerful in alternative splicing studies of annotated genes, various microarray technologies are limited by high background, cross-hybridization, low sensitivity for lowly expressed transcript variants, physical space, and reliance on previous gene and exon annotations. Tag-based profiling, such as serial analysis of gene expression (SAGE) [22], cap analysis of gene expression (CAGE) [23], and polony multiplex analysis of gene expression (PMAGE) [24] address these limitations by detecting lowly expressed and novel gene transcripts. However, these methods are incredibly time-consuming and expensive.

Recently developed high-throughput sequencing technologies are being used as an alternative to microarrays as a result of the relatively low-background, high sensitivity, and lack of dependence on gene and exon annotations. Sequencing technologies can produce gigabytes of nucleotide resolution data through massively parallel sequencing within a few days and for a fraction of the cost compared to conventional sequencing methods. Whole transcriptome shotgun sequencing, or RNA-seq, has become a useful technique for qualitative and quantitative identification of novel transcripts and splice variant detection more reliably than microarray technologies [4].

RNA-seq involves sequencing of cDNA reverse transcribed from total RNA of a cell population (Fig. **3B**). Cloonan *et al.*, were the first to demonstrate the power of this approach for studying the complexity and sequence content of the stem cell transcriptome [25]. By comparing sequenced total mRNA from mouse embryonic stem cells and embryoid bodies, they identified many known and novel alternative splicing events important for pluripotency and differentiation. Wang *et al.*, also demonstrated the power of this technique as they surveyed the transcriptome across ten diverse human tissues, revealing many tissue-specific alternative splicing events [5].

High-throughput sequence based methods have also been developed to identify regulation of alternative splicing by various RBPs. Although only a few splicing factors have been demonstrated to be important in determining the pluripotency and differentiation state of stem cells, it has been suggested that many protein factors are involved in the accurate regulation of alternative splicing in stem cell biology. It is believed that all splicing events are regulated by specific splicing factors that interact directly with pre-mRNA. Thus, identifying the specific targets of these RBPs and their mechanism of action is necessary as we unravel the rules for alternative splicing in stem cell differentiation.

The Darnell laboratory developed a method called cross-linking immunoprecipitation (CLIP), which exposes the cells to UV crosslinking to irreversibly cement physical RNA-protein interaction [26]. We, and others, have optimized the CLIP method to be coupled with high-throughput sequencing of the bound RNA to allow for the large-scale identification of endogenous RBP targets *in vivo* [27-30] (Fig. 3C). This high-throughput experimental approach, referred to as CLIP-seq, or HITS-CLIP, begins with the stabilization of *in vivo* protein-RNA interactions with UV radiation. This is followed by immunoprecipitation of the protein of interest with a specific antibody, RNA trimming by a nuclease, application of a 3' linker and 5' radio-label, and isolation of the protein-mRNA complex by SDS-PAGE. Short RNA sequences 50-100 nucleotides long shielded by the crosslinked protein of interest are then amplified by RT-PCR and sequenced on high throughput sequencing instruments.

This technique allows for a genome-wide analysis of targets for a specific RBP in the genome. Analysis of these targets can reveal sequence-level understanding of an RBPs specific role in splicing regulation. CLIP-seq has been successful in providing an RNA-map of targets, and delineating alternative splicing events regulated by NOVA [29], two SR proteins, SF2/ASF [27] and SFRS1 [28], and FOX2 in stem cells [30]. RNA-maps are a useful tool in the elucidation of the splicing code. A handful of RBPs have been strongly linked to alternative splicing and continued efforts are rapidly identifying the other functional elements of this highly regulated process.

BIOINFORMATIC ANALYSIS FOR DETECTING ALTERNATIVE SPLICING AND SRES

The vast amounts of data generated by high-throughput sequencing necessitate efficient computational approaches to analyze and interpret the data. Traditionally, alternative splicing events and SREs within the transcriptome are studied using computational analyses of publicly available expressed sequence tag (EST) databases. However, ESTs are limited by their 3' or 5' end bias, their disproportionate weight of well-studied tissues and conditions, and their insensitivity among paralogs. SELEX (systematic evolution of ligands by exponential enrichment) is a experimental method that is often complimented with *in silico* approaches for indentifying SREs [31] and has been successful in identifying ESEs for SR proteins [32, 33]. Bioinformatic analysis of sequence data for transcripts identified by microarray and sequencing experiments will be vital for detecting alternative splicing events and unraveling the SREs that mediate those events.

The utility of microarray data relies on computational post-processing to remove background and quantify differential expression. Algorithms such as SPLICE [34], and ANOSVA [35] have been developed to detect differentially spliced exons from sets of microarray data. These techniques rely on statistical tests of significance and have proven to be accurate and useful. However, they are not optimized for novel splice variant detection, nor can they accurately quantify expression of isoforms from a single pre-mRNA transcript due to low sensitivity of the statistics and inherent microarray limitations [36]. In addition to locating differentially expressed exons, global bioinformatic approaches are necessary to examine the neighboring sequences of these exons for discovery of SREs. In our microarray-based study of alternative splicing, we uncovered a conserved 'GCAUG' *FOX1/2* binding motif near splice sites for exons with differential expression in neuronal differentiation [14].

More recently, microarray technology has been overpowered by the advantages of high-throughput sequencing data, and thus many have focused their attention toward generating algorithms to analyze this source of data. In order to render this wealth of sequencing data useful, the resultant reads must first be pre-processed to remove PCR and sequencing artifacts and then mapped to a reference genome. Due to repetitive regions of the genome, and the size and error rate of the reads output by high-throughput sequencing devices (~35-100 nucleotides), uniquely mapping the reads to known genes has become a challenge. The large structural variations caused by alternative splicing events and other post-transcriptional modifications enormously complicate the sequence mapping problem, since reads are capable of mapping across novel and canonical exon-exon junctions.

Many tools are rapidly being developed to approach this bottleneck of accurately mapping short reads to the transcriptome. Tools such as BLAT [37], Bowtie [38], Maq [39], ELAND (Illumina), and BFAST [40] use different algorithms to quickly align millions of reads to a reference genome. Algorithms allow for some mismatches, and have the option to consider the sequence quality values of each read for improved mapping. These techniques are useful but often leave many reads unmapped [5].

Mapping approaches specific for alternative splicing detection have also been developed. Since sequenced reads come from the highly-processed transcriptome, many reads do not map directly to the reference genome because many reads do not fall entirely within exons or introns and span exon-exon junctions. Reads that do not map directly to the genome can be mapped using standard mapping algorithms to custom-generated exon-exon junction databases [5, 25, 41, 42]. In most cases, these approaches are known to still leave one-third of the reads unmappable [5]. This may be due to non-unique repetitive regions in the genome, sequencing errors, or reads spanning novel splice junctions. Other tools, such as TOPHAT [43], attempt to tackle the alternative splicing mapping problem directly, by mapping reads to novel splice junctions defined significant within the data. Continual improvement of these mapping tools will be necessary for accurate analysis of alternative splicing within high-throughput sequencing data.

Part of the power of algorithmic approaches is their ability to identify sequence patterns such as those due to conservation, evolution, and enrichment. Computational approaches have been surprisingly successful in defining *cis*-regulatory elements within the genome [3]. For example, RESCUE-ESE (Relative Enhancer and Silencer

Classification by Unanimous Enrichment of Exon Splicing Enhancers) uses statistical analysis of exon-intron and splice site composition to predict the ESE activity of sequences [44]. Another statistical method, called CoCOA (orthogonalized co-occurrence analysis) identifies enriched, co-occurring oligonucleotide sequence pairs in exon flanking regions that may cooperate during splicing [45]. With enough knowledge of key functional elements, purely computational methods can be accurately applied to predict alternative splicing events.

COMBINATORIAL SPLICING PREDICTION

As more and more factors that influence splicing are revealed by microarray and sequencing technologies, the splicing code continues to unravel. The splicing field is now in cataloging phase, as researchers attempt to pull together the individual regulatory elements involved in constitutive and alternative exon splicing into a cohesive code. Various computational approaches are being employed to sort and combine identified SREs as part of a layered splicing network. Once identified, sequence elements and traits near alternatively spliced exons provide a set of features useful for training machine learning classifiers to predict exon inclusion or exclusion. The successes of algorithms like ExonScan, which uses splice site models and known ESEs and ESSs to predict the splicing patterns of transcripts [46], prove that this sort of predictive analysis of alternative splicing is possible.

Other algorithms that rely upon basic splicing elements such as sequence conservation and motifs for their predictions have been successful in predicting unannotated, alternative exons. Sorek *et al.,* were the first to make alternative splicing predictions without the use of ESTs and were able to determine many novel splice variants using only the surrounding genomic sequence data [47]. Another algorithm, ACESCAN, combined multiple sequence features into a regularized least-squares classifier to identify exons subject to evolutionarily conserved exon skipping [48]. This method accurately predicts about 2,000 alternative conserved exons in the human and mouse genomes. In 2005, Dror *et al.,* used a support vector machine (SVM) classifier with seven features including, length of the exon, exon divisibility by three, conservation to mouse, and various flanking sequence features to predict alternatively spliced exons [49]. While these predictive methods combine multiple splicing features, many more features exist, and can be combined for improved predictive algorithms.

Brendan Frey's group has recently developed the first combinatorial method for predicting tissue-specific exon inclusion, that combines over 1000 established *cis* and *trans* SREs into a bayesian network classifier [50]. This approach delineates tissue-specific splice patterns and can correctly predict splicing for more than 60% of cassette exons across 4 mouse tissue types. This approach supports a giant leap in our ability to predict the occurrence of tissue-specific alternative splicing.

CONCLUSIONS AND FUTURE PERSPECTIVES

Post-transcriptional gene expression regulation, specifically through alternative splicing, is crucial for most cell processes. While alternative splicing has only recently been implicated in stem cell biology, it is clear that alternative splicing plays significant roles in fine-tuning the proteome during lineage specification. Novel tissue-specific transcript variants are quickly being discovered through microarray and sequencing technologies, and techniques such as CLIP-seq are rapidly contributing to our understanding splicing code's defining factors. Future studies will rely more heavily upon computational tools that draw increasingly significant results from this abundance of data. Eventually, understanding of alternative splicing may guide new ways to control stem cell fate, allow finer distinction among cell subtypes, and provide a molecular basis for human disease model development.

CONFLICT OF INTEREST

None declared.

ACKNOWLEDGEMENTS

The authors thank K.R. Hutt, and D.A. Nelles for critical reading of the review. S. Huelga is supported by the Cancer Cell Biology Training Grant Program at the University of California, San Diego. Our work is supported by grants from the US National Institutes of Health (HG004659 and GM084317), and the Stem Cell Program at the University of California, San Diego.

REFERENCES

[1]   Martinez-Contreras R, Cloutier P, Shkreta L, *et al.* hnRNP proteins and splicing control. Adv Exp Med Biol 2007; 623:123-47.

[2]   Mayeda A, Helfman DM, Krainer AR. Modulation of exon skipping and inclusion by heterogeneous nuclear ribonucleoprotein A1 and pre-mRNA splicing factor SF2/ASF. Mol Cell Biol 1993; 13 (5):2993-3001.

[3]   Wang Z, Burge CB. Splicing regulation: from a parts list of regulatory elements to an integrated splicing code. RNA 2008; 14 (5):802-13.

[4]   Pan Q, Shai O, Lee LJ, Frey BJ, Blencowe BJ. Deep surveying of alternative splicing complexity in the human transcriptome by high-throughput sequencing. Nat Genet 2008; 40 (12):1413-5.

[5]   Wang ET, Sandberg R, Luo S, *et al.* Alternative isoform regulation in human tissue transcriptomes. Nature 2008; 456 (7221):470-6.

[6]   Takahashi K, Tanabe K, Ohnuki M, *et al.* Induction of pluripotent stem cells from adult human fibroblasts by defined factors. Cell 2007; 131 (5):861-72.

[7]   Cauffman G, Liebaers I, Van Steirteghem A, Van de Velde H. POU5F1 isoforms show different expression patterns in human embryonic stem cells and preimplantation embryos. Stem Cells 2006; 24 (12):2685-91.

[8]   Lee J, Kim HK, Rho JY, Han YM, Kim J. The human OCT-4 isoforms differ in their ability to confer self-renewal. J Biol Chem 2006; 281 (44):33554-65.

[9]   Atlasi Y, Mowla SJ, Ziaee SA, Gokhale PJ, Andrews PW. OCT4 spliced variants are differentially expressed in human pluripotent and nonpluripotent cells. Stem Cells 2008; 26 (12):3068-74.

[10]   Ates K, Yang SY, Orrell RW, *et al.* The IGF-I splice variant MGF increases progenitor cells in ALS, dystrophic, and normal muscle. FEBS Lett 2007; 581 (14):2727-32.

[11]   Gopalakrishnan S, Van Emburgh BO, Shan J, *et al.* A novel DNMT3B splice variant expressed in tumor and pluripotent cells modulates genomic DNA methylation patterns and displays altered DNA binding. Mol Cancer Res 2009; 7 (10):1622-34.

[12]   Patel NA, Song SS, Cooper DR. PKCdelta alternatively spliced isoforms modulate cellular apoptosis in retinoic acid-induced differentiation of human NT2 cells and mouse embryonic stem cells. Gene Expr 2006; 13 (2):73-84.

[13]   Lin H, Shabbir A, Molnar M, *et al.* Adenoviral expression of vascular endothelial growth factor splice variants differentially regulate bone marrow-derived mesenchymal stem cells. J Cell Physiol 2008; 216 (2):458-68.

[14]   Yeo GW, Xu X, Liang TY, *et al.* Alternative splicing events identified in human embryonic stem cells and neural progenitors. PLoS Comput Biol 2007; 3 (10):1951-67.

[15]   Salomonis N, Nelson B, Vranizan K, *et al.* Alternative splicing in the differentiation of human embryonic stem cells into cardiac precursors. PLoS Comput Biol 2009; 5 (11):e1000553.

[16]   Johnson JM, Castle J, Garrett-Engele P, *et al.* Genome-wide survey of human alternative pre-mRNA splicing with exon junction microarrays. Science 2003; 302 (5653):2141-4.

[17]   Pan Q, Shai O, Misquitta C, *et al.* Revealing global regulatory features of mammalian alternative splicing using a quantitative microarray platform. Mol Cell 2004; 16 (6):929-41.

[18]   Blanchette M, Green RE, Brenner SE, Rio DC. Global analysis of positive and negative pre-mRNA splicing regulators in Drosophila. Genes Dev 2005; 19 (11):1306-14.

[19]   Zhang Z, Hesselberth JR, Fields S. Genome-wide identification of spliced introns using a tiling microarray. Genome Res 2007; 17 (4):503-9.

[20]   Gardina PJ, Clark TA, Shimada B, *et al.* Alternative splicing and differential gene expression in colon cancer detected by a whole genome exon array. BMC Genomics 2006; 7:325.

[21]   Keene JD, Komisarow JM, Friedersdorf MB. RIP-Chip: the isolation and identification of mRNAs, microRNAs and protein components of ribonucleoprotein complexes from cell extracts. Nat Protoc 2006; 1 (1):302-7.

[22]   Velculescu VE, Zhang L, Vogelstein B, Kinzler KW. Serial analysis of gene expression. Science 1995; 270 (5235):484-7.

[23]   Shiraki T, Kondo S, Katayama S, *et al.* Cap analysis gene expression for high-throughput analysis of transcriptional starting point and identification of promoter usage. Proc Natl Acad Sci USA 2003; 100 (26):15776-81.

[24]   Kim JB, Porreca GJ, Song L, *et al.* Polony multiplex analysis of gene expression (PMAGE) in mouse hypertrophic cardiomyopathy. Science 2007; 316 (5830):1481-4.

[25]   Cloonan N, Forrest AR, Kolle G, *et al.* Stem cell transcriptome profiling *via* massive-scale mRNA sequencing. Nat Methods 2008; 5 (7):613-9.

[26]   Ule J, Jensen KB, Ruggiu M, *et al.* CLIP identifies Nova-regulated RNA networks in the brain. Science 2003; 302 (5648):1212-5.

[27]   Sanford JR, Coutinho P, Hackett JA, *et al.* Identification of nuclear and cytoplasmic mRNA targets for the shuttling protein SF2/ASF. PLoS One 2008; 3 (10):e3369.

[28]   Sanford JR, Wang X, Mort M, *et al.* Splicing factor SFRS1 recognizes a functionally diverse landscape of RNA transcripts. Genome Res 2009; 19 (3):381-94.

[29]   Licatalosi DD, Mele A, Fak JJ, *et al.* HITS-CLIP yields genome-wide insights into brain alternative RNA processing. Nature 2008; 456 (7221):464-9.

[30]   Yeo GW, Coufal NG, Liang TY, *et al.* An RNA code for the FOX2 splicing regulator revealed by mapping RNA-protein interactions in stem cells. Nat Struct Mol Biol 2009; 16 (2):130-7.

[31]   Lorenz C, von Pelchrzim F, Schroeder R. Genomic systematic evolution of ligands by exponential enrichment (Genomic SELEX) for the identification of protein-binding RNAs independent of their expression levels. Nat Protoc 2006; 1 (5):2204-12.

[32]   Tacke R, Manley JL. The human splicing factors ASF/SF2 and SC35 possess distinct, functionally significant RNA binding specificities. EMBO J 1995; 14 (14):3540-51.

[33]   Cavaloc Y, Bourgeois CF, Kister L, Stevenin J. The splicing factors 9G8 and SRp20 transactivate splicing through different and specific enhancers. RNA 1999; 5 (3):468-83.

[34]   Hu GK, Madore SJ, Moldover B, *et al.* Predicting splice variant from DNA chip expression data. Genome Res 2001; 11 (7):1237-45.

[35]   Cline MS, Blume J, Cawley S, *et al.* ANOSVA: a statistical method for detecting splice variation from expression data. Bioinformatics 2005; 21 Suppl 1:i107-15.

[36]   Cuperlovic-Culf M, Belacel N, Culf AS, Ouellette RJ. Data analysis of alternative splicing microarrays. Drug Discov Today 2006; 11 (21-22):983-90.

[37]   Kent WJ. BLAT--the BLAST-like alignment tool. Genome Res 2002; 12 (4):656-64.

[38]   Langmead B, Trapnell C, Pop M, Salzberg SL. Ultrafast and memory-efficient alignment of short DNA sequences to the human genome. Genome Biol 2009; 10 (3):R25.

[39]   Li H, Ruan J, Durbin R. Mapping short DNA sequencing reads and calling variants using mapping quality scores. Genome Res 2008; 18 (11):1851-8.

[40]   Homer N, Merriman B, Nelson SF. BFAST: an alignment tool for large scale genome resequencing. PLoS One 2009; 4 (11):e7767.

[41]   Wang L, Xi Y, Yu J, *et al.* A statistical method for the detection of alternative splicing using RNA-seq. PLoS One 2010; 5 (1):e8529.

[42]   Li H, Lovci MT, Kwon YS, *et al.* Determination of tag density required for digital transcriptome analysis: application to an androgen-sensitive prostate cancer model. Proc Natl Acad Sci USA 2008; 105 (51):20179-84.

[43]   Trapnell C, Pachter L, Salzberg SL. TopHat: discovering splice junctions with RNA-Seq. Bioinformatics 2009; 25 (9):1105-11.

[44]   Fairbrother WG, Yeh RF, Sharp PA, Burge CB. Predictive identification of exonic splicing enhancers in human genes. Science 2002; 297 (5583):1007-13.

[45]   Friedman BA, Stadler MB, Shomron N, Ding Y, Burge CB. Ab initio identification of functionally interacting pairs of cis-regulatory elements. Genome Res 2008; 18 (10):1643-51.

[46]   Wang Z, Rolish ME, Yeo G, *et al.* Systematic identification and analysis of exonic splicing silencers. Cell 2004; 119 (6):831-45.

[47]   Sorek R, Shemesh R, Cohen Y, *et al.* A non-EST-based method for exon-skipping prediction. Genome Res 2004; 14 (8):1617-23.

[48]   Yeo GW, Van Nostrand E, Holste D, Poggio T, Burge CB. Identification and analysis of alternative splicing events conserved in human and mouse. Proc Natl Acad Sci USA 2005; 102 (8):2850-5.

[49]   Dror G, Sorek R, Shamir R. Accurate identification of alternatively spliced exons using support vector machine. Bioinformatics 2005; 21 (7):897-901.

[50]   Barash Y, Calarco JA, Gao W, *et al.* Deciphering the splicing code. Nature 2010; 465: 53-59.

CHAPTER 9

Computational Biology of microRNA-Pluripotency Gene Networks in Embryonic Stem Cells

Preethi H. Gunaratne[1,2,3,*] and Jayantha B. Tennakoon[1]

[1]Department of Biology & Biochemistry, University of Houston; [2]Department of Pathology and [3]Human Genome Sequencing Center, Baylor College of Medicine, Houston

Abstract: Spectacular advances in technology and computational biology spawned in large part by the Human Genome Project have been instrumental in transforming the field of embryonic stem cell biology. From the findings reported in the last decade it is clear that properties unique to embryonic stem cells (ESCs) are regulated not by individual genes but by complex gene networks that include both genes and ~22 nt noncoding microRNAs that act to integrate multiple genes across diverse signaling pathways to regulate self-renewal and differentiation. In this chapter we will discuss the evolution of our understanding of regulatory networks underlying stem cell self-renewal and pluripotency made possible through highthroughput genomic studies. In the last decade molecular technologies that revealed key transcription and epigenetic factors in ESCs have given way to highthroughput microarray and Next Generation Sequencing technologies. These large-scale genomics datasets analyzed through the latest bioinformatic and computational methods have been instrumental in transforming the field of embryonic stem cells. We will trace the history of ES cells to briefly discuss key genes and microRNAs that have been established to regulate self-renewal and pluripotency in mouse and human prior to the genomics revolution. We will then discuss the latest technologies and computational algorithms that have been instrumental in revealing genome-wide changes associated with self-renewal and differentiation at the genetic and epigenetic levels to yield the current systems-level understanding of embryonic stem cells.

Keywords: miRNA, stem cells, pluripotency, epigenome, reprogramming, microarrays, gene networks, Next generation sequencing (NGS), Chromatin immunoprecipitation (ChIP), target prediction, iPSCs, bioinformatics, differentiation, polycomb group proteins (PcGs), methylation, transcription factors.

Oct4, Sox2, NANOG, THE TRINITY OF TRANSCRIPTION FACTORS ARE ESSENTIAL FOR EMBRYONIC STEM CELL SELF-RENEWAL AND PLURIPOTENCY

During their life history mammalian embryonic stem cells transition from primarily symmetrically dividing cells to mostly asymmetrically dividing cells

*Address correspondence to P. H. Gunaratne: Department of Biology & Biochemistry, University of Houston and Department of Pathology, Human Genome Sequencing Center, Baylor College of Medicine, Houston, TX 77204; E-mail: preethig@bcm.tmc.edu

during late to mid-gestation [1]. During this period the totipotent embryonic stem cells systematically restrict their potential through a series of differentiation steps. The first lineage restriction event generates pluripotent embryonic stem cells (ESCs) can form the inner cell mass (ICM) and the blastocyst and trophoblastic stem (TSCs) cells that define the extraembryonic lineage including trophoectoderm, the primordial endoderm and visceral endoderm [1]. Cells in the epiblast then further differentiate to generate multipotent tissue progenitors that can form all of the cell types in a specific lineage. Terminally differentiated cells that are typically entirely restricted in their potential make up most of the cells in our body. The key genetic and epigenetic factors that are necessary and sufficient for this carefully controlled perfectly orchestrated process are now quite well understood. Most of our understanding of these processes come from studying mouse embryonic stem cells (ES) that are derived from the inner cell mass (ICM) of the mammalian blastocyst. ES cell pluripotency and self-renewal are controlled by a network of transcription factors including Oct4, Sox2 and NANOG [2, 3]. These transcription factors in conjunction with polycomb group (PcG) repressor proteins and histone/DNA methyl transferases establish dynamic patterns of 'bivalent chromatin' consisting of repressive H3K27me3 and permissive H3K4me3 histone modifications in the promoters of several hundred differentiation genes [3]. This bi-valent state defines a fluid epigenome that can easily transition between self-renewal and differentiation in response to various intra and extracellular cues [2, 3].

TERMINALLY DIFFERENTIATED CELLS CAN BE REPROGRAMMED TO GENERATE IPSCSOR BE TRANS-DIFFERENTIATED INTO OTHER CELL TYPES

The first stable pluripotent hESC was established by Thompson *et al.,* 1998 [4]. Human ES cells (hES) and mouse ES cells (ES) have some striking differences. The LIF pathway supports the maintenance of mouse ES cells but not hES. BMPs which support self-renewal in ES mediate differentiation of hES [5]. FGF and TGFβ/Activin/Nodal signaling plays a central role in hES self-renewal [5].

Recently, ectopic expression of a handful of genes Oct4, Sox2, Klf4, cMyc and Oct4/Sox2/NANOG/LIN28 were shown to be necessary and sufficient to re-programme murine embryonic fibroblast (MEF) cells into induced pluripotent stem cells (iPSCs) which can form viable late-term embryos [6-8]. Other work suggests that LIN28 and cMYC may be optional [9]. These studies imply that epigenomes of terminally differentiated cells also have some fluidity that can be manipulated to both generate iPSCs through reprogramming or to transdifferentiate into other cell types [10, 11]. The complexity of gene networks underlying stem cell properties is exemplified by the fact that current strategies for generating iPSCs are hampered by low efficiencies and a 30% chance of forming invasive teratomas [12]. Recently, very exciting work by Judson *et al.,* 2009 revealed that a single miRNA, miR-294 was able to replace cMYC [13]. The most striking observation to emerge from this work is that by contrast to very low yields of iPSCs and ~30% invasive teratomas the Oct4, Sox2, Klf4, miR-294 combination yielded 100% iPSCs and 0% invasive teratomas suggesting that incorporating miRNAs into current iPSC strategies will significantly enhance the efficacy and safety [13, 14].

GENOME-WIDE PROMOTER OCCUPATION STUDIES FOR Oct4, Sox, NANOG AND THE PcG REPRESSORS PRC1 AND PRC2 REVEAL CORE CIRCUITRY REGULATING ES CELL PROPERTIES

The regulatory circuits governed by Oct4, Sox2 and NANOG in human ES cells were first determined by Boyer *et al.,* 2005 using DNA microarrays carried out on chromatin immunoprecipitated DNA using antibodies to each of the three transcription factors. Two other studies that followed by Boyer *et al.,* 2006 and Lee *et al.,* 2006 Identified the regulatory circuits governed by Polycomb Group Repressor (PcG) complexes in mouse and human ES cells respectively [2, 3]. In the first study it was found that 2260 genes in the human genome were occupied by at least one of the three transcription factor. Approximately 60% of the genes were transcribed and 40% were silent [2]. A large number of homeodomain genes were among the silent group [5]. All three transcription factors were found to co-localize and form feed-forward loops at 353 genes and 2 miRNA promoters [2]. In the second study PcG complexes PRC1 and PRC2 were found to co-localize at the promoters of 512 genes in mouse ES cells [15]. The majority of genes exhibiting promoter co-occupation were developmental genes and all of their promoters exhibited H3K27me3 which has been established to be a classic mark for PRC2-mediated repression [13].

MicroRNAs PLAY IMPORTANT ROLES IN STEM CELL SELF-RENEWAL AND DIFFERENTIATION

Several lines of evidence suggest that miRNAs are likely to play important roles in the transition during both embryonic development and retinoic acid (RA)-induced differentiation of ES cells [16]. *Dicer*, which is essential for miRNA maturation, is critical for self-renewal and pluripotence. *Dicer*$^{-/-}$ ES cells are characterized by defects in proliferation, miRNA maturation, and failure to differentiate [17]. ES *Dicer*$^{-/-}$ cells fail to repress *Oct4* expression and are impaired in their ability to differentiate [17]. MicroRNAs (miRNAs) have recently exploded in the scene as important and widespread agents of gene silencing of protein coding genes [18, 19]. Previously assumed to represent 'junk' due to their inability to make protein the discovery of miRNAs and their role in RNA-interference has caused a paradigm shift and brought them to a central and pivotal position in biology today. MicroRNAs are ~22-nucleotide single-stranded RNAs that are critical elements of gene regulatory networks in plants and animals [18, 19]. These small RNAs are encoded in the genome and transcribed by RNA polymerase II as large primary microRNAs (pri-miRNAs). The pri-miRNAs are processed into short stem loop structures (pre-miRNA) by DROSHA, an RNase III enzyme [16]. Export of the pre-microRNA into the cytoplasm is directed by the Ran-GTP-dependent transporter EXPORTIN 5, and the pre-microRNA is subsequently cleaved by a second RNase III, DICER, to excise the ~22-nucleotide microRNA:microRNA* duplex. One strand of the duplex is preferentially incorporated into the RNA-induced silencing complex (RISC), ultimately catalyzing the degradation and/or translational repression of complementary target mRNAs. Most animal microRNAs are thought to bind sequences of imperfect complementarily within 3' untranslated regions (3' UTR) of individual mRNAs, leading to translational repression of target transcripts. A single miRNA is predicted to target and silence several hundred protein coding genes through complementary base pairing with sequences in the 3'-untranslated regions of protein coding transcripts (3'-UTRs). A single protein coding gene has binding sites for several miRNAs. Decisive roles for microRNAs have been described across mammalian development, including but not limited to cardiogenesis, hematopoiesis, and HOX-mediated limb patterning [20]. These functions depend on microRNA regulation of fundamental cellular processes, namely growth, differentiation, and apoptosis. Over 75% of miRNAs in ES cells are represented by 6 clusters in the mouse genome [21]. These include the mmu-mir-290-295 cluster, miR-467-cluster, miR-17-92 cluster, Chromosome 12 cluster, mmu-mir-15a, b, and mmu-mir-21 [21].We and others have determined the miRNAome of ES [16] and hES [22].

IN SILICO SEARCH FOR CONSERVED SMALL RNAs

As is often the case at a scientific frontier, much can be learned by the study of a specific group of functional elements in our genome through the synergy of experimental data and *in silico* predictions. For example, Ambros [18] identification of many transcripts in worm, anti-sense to protein-coding mRNA, all of a characteristic length, suggests that a general mechanism may be operative; identification of only one or two examples would be much harder to interpret. Similarly, the identification of many short transcripts that exhibit a characteristic length, secondary structure and/or partial complementarity to specific locations in the genome could indicate so-far undiscovered mechanisms for the function of short RNAs. Indeed, the most successful efforts at inferring targets have been based on statistical study of the set of confirmed microRNAs as a whole. In plants, this study was decisive even though it relied on a relatively small number of microRNA sequences. In animals, the mechanism is much more subtle, and a comprehensive account of distinct microRNA sequences becomes even more vital.

In recent years, short RNAs have been found to function in a range of processes, ranging from imprinting and epigenesis, developmental timing, suppression of viral replication *via* silencing, and numerous diseases including cancer. For none of these roles do there exist reliable estimates of the total number of distinct short RNA sequences that function in a given capacity. In general, the more sensitive the tools applied, the richer the yield of sequence and modalities for short RNA that are revealed.Increasingly, bioinformatics has been used to screen genomes for specific transcripts that have eluded discovery. The motivation for this approach is based on the premise that functional constraints on sequence variation during evolution can be exploited to infer functional genomic elements. This premise can be used to find novel short RNA because many known miRNAs are strongly and deeply conserved. A number of different miRNA prediction algorithms, most of them relying on synteny or whole-genome alignment, have been developed for genome-wide searches. Some commonly used tools are MiRseeker [23] and MiRscan [24]. Dozens of new miRNAs identified by these methods have subsequently been experimentally confirmed. A more recent and sophisticated approach involves "phylogenetic shadowing', wherein comparison of known miRNA genes by multiple whole-genome alignment of closely related genomes is used to derive a conservation profile for miRNA [25].

INTEGRATION OF MicroRNAs INTO ES CELL GENE NETWORKS USING MICROARRAY ANALYSIS

Microarray Platforms

High throughput analysis of genome-wide microRNA, gene and transcription factor binding and epigenetic modification can be assessed through microarrays. In this technique fluorescently labeled cDNA or chromatin immunoprecipiated DNA is used to probe oligonucleotides immobilized on a solid platform, flourescent signals resulting from the hybridization reaction are read by a laser scanner and analyzed using computer software [26].

Microarrays are primarily available in three different varieties including tiling arrays that contain probes for entire chromosomes or genomes used for ChIP-chp experiments to measure genome-wide transcription factor binding, histone modification or methylation. Gene expression studies can be done with cDNA microarrays or gene expression arrays containing multiple probes for different exons of each RefSeq gene. More recently a number of different miRNA microarray platforms have been developed to assess genome-wide microRNA expression. Some commonly used platforms include Affymetrix GeneChip [27] (http://www.affymetrix.com), Agilent microarrays [28] (http://www.agilent.com), [28] Nimblegen [29] (http://www.nimblegen.com/), Combimatrix [30] (http://www.combimatrix.com) and Illumina [31] bead arrays (http://www.illumina.com).

MicroRNA Microarrays

Designing miRNA microarrays is somewhat challenging since probes designed to detect the mature miRNA sequence also hybridizes to the precursor hairpin miRNA. Furthermore, the wide range of melting temperatures of the mature miRNAs in miRBase could cause non uniform hybridiazation and erratic signals [32]. The different platforms used by manufacturers therefore often use chemically modified probes to circumvent these challenges. Novel classes of conformationally restricted oligonucelotide analogs with high thermal stabilities termed locked nucleic acids (LNA), and incorporation of chemically modified A/T analogues to match hybridization affinity of G/C analogues of oligonucleotides are commonly used for miRNA array design [32-36]. Agilent Exiqon, Invtirogen and LC Sciences are examples of commonly used systems for microRNA detection through microarray.

One of the first genome-wide analyses of non-coding RNAs that are regulated in ES cells during self-renewal and differentiation was carried out by Gu *et al.,* 2008 [16]. Using a bioinformatics program that was designed to carry out an *in silico* search for micro-conserved elements (MCE) they uncovered over 500 short non-coding small RNAs that are expressed in adult tissue progenitors and highly enriched in ES cells and rapidly down regulated upon retinoic acid induced differentiation. Only 20% of these small RNAs were Dicer-dependent and the majority failed to down-regulate in GCNF$^{-/-}$ mutants. The orphan nuclear receptor GCNF (germ cell nuclear factor) is a transcriptional repressor of both Oct4 and Nanog [1]. ES [GCNF$^{-/-}$] and ES [Dicer$^{-/-}$] are both characterized by failure to repress Oct4 expression and impaired differentiation [37]. Gu *et al.,* 2008 reported that the temporal patterns of many mRNAs expressed during RA-induced differentiation in ES cells are inversely correlated with the expression of the miRNAs that regulate them. They also found that the subset of expression patterns that correlate with the phenotypes of pluripotence, self-renewal and differentiation were significantly altered in GCNF$^{-/-}$ ES cells. In addition they found a large number of micro-conserved non-coding RNAs that are highly enriched in ES cells and strongly down-regulated upon differentiation. These MCE-non-coding RNAs are also expressed in tissue progenitors and adult stem cells from the hematopoietic and hepatic systems. Their work suggests that miRNAs and other noncoding RNAs are highly regulated during ES cell self-renewal and differentiation [16].

INTEGRATION OF MicroRNAs INTO ES CELL GENE NETWORKS USING ChIP-SEQUENCING AND SMALL RNA SEQUENCING

Whole Genome Arrays – ChIP-ChIP

Genome-wide protein transcription factor binding is now a necessary layer for understanding gene networks underlying complex biological processes such as embryonic stem cell self-renewal and pluripotency. In the past this was achieved by cross linking proteins to associated DNA by formaldehyde treatment or UV induction followed by shearing of DNA and purifying DNA bound to specific transcription factors through immunoprecipitation (ChIP) using antibodies to the transcription factor of interest. These chromatin immunoprecipitated DNA was released through reverse cross linking and the transcription factor bound DNA was typically analyzed through PCR (ChIP-PCR) for single gene analysis. Whole genome studies were done by using the ChIPed DNA to probe microarrays containing either all known promoters in the genome of interest or more recently whole chromosome or genome tiling arrays. The design of the oligonucleotide-based whole genome array set and data extraction methods are described in Lee *et al.*, 2006 [15]. Using microarrays from Agilent Technologies (http://www.agilent.com) Marson *et al.*, 2008 [38] used procedures described by Lee *et al.*, [15] 2006 to examine genome-wide co-occupation of the pluripotency factors Oct4, Nanog and Sox2.

ChIP-Sequencing

Recently developed 'Next Generation Sequencing' (NGS) technologies have been extremely useful in determining genome-wide miRNA expression and histone methylation patterns through small RNA sequencing and Chromatin ImmunoPrecipitation Sequencing (ChIPSeq) [39]. With the development of Next Generation Sequencing technologies our ability to interrogate genome-wide transcription factor binding sites was transformed. Purified ChIP DNA is fragmented to generate 150-300 nucleotide fragments, end repaired and ligated to adaptors and subjected to PCR amplification [38]. Amplified fragments are gel purified and the extracted DNA is massively amplified using bridge amplification and cluster generation on a Illumina/Solexa flow-cell and then subjected to 26-70 bases sequencing according to standard protocols optimized by Illumina.

Following sequencing images are extracted and sequences are aligned to reference genomes of choice.

Small RNA Sequencing [40]

The overall procedure of small RNA sample preparation involves ligating sets of proprietary adaptor sequences at the 5' and 3' ends of extracted RNA samples. Adaptor ligated RNA samples are reverse transcribed and thereafter PCR amplified for enrichment. In order to extract the amplified fraction corresponding to small RNAs samples are run on acrylamide gels from which the necessary molecules corresponding to a particular molecular weight could be excised. The excised fraction is further amplified on a solid platform using a PCR reaction where small DNA fragments having conjugate base pairs complementary to ligated adaptors, function as primers immobilized on a solid platform. During the course of amplification DNA molecules attached to the solid platform at one end due to complementarity of ligated adaptors to immobilized oligonucleotides bend over and form arch or bridge like duplex DNA structures by attaching to adjacent adapters at the other end on the platform. Since characteristic arch like structures are formed during the process, this step is often referred to as "bridge amplification". The bridge-amplified samples on the platform are finally introduced to the Illumina GA-2 genome analyzer where a final PCR stage takes place using fluorescently labeled nucleotides. This step is particularly interesting for the fact that at each occurrence where a nucleotide gets incorporated the polymerase reaction is terminated and the full gamut of four color coded nucleotides incorporated at that time point are photographed using a smart camera. Since the termination step is one that is reversible, the polymerase reaction is reinitiated after photography, thus a new set of labeled nucleotide incorporation takes place facilitating numerous subsequent cycles of reversible termination followed by digital photography where data is stored in a computer. Thereafter using sophisticated algorithms the color-coded photographs are de-convoluted to corresponding base pairs, which can be finally assembled to short sequences representing the original RNA molecules in the sample. Streamlined protocols assisted with advanced algorithms ensure that the final assembly is built at a non-saturated PCR phase dataset so that the quantitative data representing different sequences would be proportionate to the actual numbers of sequences found in the original samples. A detailed protocol for carrying out sample preparation for small RNA sequencing using the Illumina GA-2 sequencing platform can be found at the following link which leads to the Illumina small RNA preparation kit resource page at the manufacturers website (http://www.illumina.com/products/small_rna_sample_prep_kit.ilmn)

Once the final assembly is made it is necessary to map the resultant sequences to a reference database to annotate the small RNA sequences. Numerous computer programs have been written to fulfill the task of mapping small RNA sequences, which normally range in the numbers of 5 to 10 million reads per sample per lane of the illumina solid-state platform. The programs written are largely tailor maid to fulfill the researchers requirements demanded by different applications. In a gross sense however, a well written algorithm should be capable of scanning the resultant sequences against a reference database, assess and qualify whether the reads evolved from processable precursors, then finally score and arrange the reads according to abundance and degree of matching agreement to reference sequences. Additional modifications can be incorporated into algorithms *via* source code to evaluate possible novel candidates that are not in the reference databases. Although different criteria can be incorporated to an algorithm to qualify reads as novel candidates as these would be based on sets of predicted *in silico* conditions it is imperative that the output of such computer generated data is functionally tested and evaluated in a laboratory setup before final submission to a curated database. Detailed protocols on miRNA sequence mapping and novel miRNA discovery can be found in Reid *et al.* 2008, Creighton *et al.* 2008 and Berezikov *et al.* 2002 [40-42].

miRNA Target Prediction Programs

A number of miRNA target prediction algorithms are publicly accessible through the Internet. While different target prediction programs employ different criteria in predicting the target repertoire of a given miRNA they all serve the same purpose. In effect suppose a researcher wants to decide on a set of predicted target candidates which would eventually be tested in a wet lab set up it is desirable that a set of candidates that are universally qualified using different criteria are chosen. Hence researchers often use a combination of algorithms to decide on miRNA targets which they eventually plan researching on the bench. What combination of algorithms to use would depend on criteria such as: the species, from which a particular miRNA originate, the degree of complementarity of the base pairs in the seed and anchor regions of a miRNA to mRNA, free energy values of binding and stringency criteria.

Some of the popular miRNA target prediction algorithms, which often appear in research articles, are described below.

TargetScan (http://www.targetscan.org/): The TargetScan algorithm uses predictions described by Lewis *et al.,* and is publicly accessible through the internet. This tool provides a user-friendly interphase, which provides, easy target prediction of several vertebrate species using a graphical user interphase. One of the key features of this algorithm is the use of stringency in seed pairing criteria and the conservedness of the miRNA-mRNA target pairs across vertebrate species. Apart from these two criteria in predicting target pairs this algorithm takes into account the number of target sites for a given miRNA in a given 3'UTR, stringency in seed pairing and the parameters, which influence site accessibility. An advantage of this algorithm is that in the mode where site conservation is taken into account there is an option to rank by preferential conservation instead of site context.PicTar (http://pictar.mdc-berlin.de/): The Pictar algorithm provides miRNA target predictions for a variety of clades, which include mammalian/vertebrate, fly and worm. The algorithm essentially takes conservation into account for all cases of target predictions. For at least one of the sites stringent seed pairing is defined as a prerequisite, moreover overall stability in pairing and the number of pairing sites are also taken into account. The site also provides useful links to other target prediction tools made publicly available by different authors. A disadvantage of the Pictar algorithm however is the inability of carrying out target predictions without taking site conservation into account. A detailed explanation on criteria employed in the PicTar algorithm is elaborated in Krek *et al.,* [43]. The miRanda algorithm, (http://www.microrna.org): The miRanda algorithm offers miRNA target predictions through a graphical user interface for human, mouse rat, drosophila and *C. Elegans.* This algorithm uses moderately stringent seed pairing and the total number of sites paring to a miRNA in determining predicted targets. PITA (http://genie.weizmann.ac.il/pubs/mir07/mir07_prediction.html): The PITA algorithm provides a convenient way of making miRNA target predictions for a variety of clades which include, vertebrates, worms as well as flies. While moderately stringent seed pairing is taken into consideration in the algorithm it possible to make target predictions with or without conservation. The overall stability of the formed miRNA-mRNA duplex structures and site accessibility are other features incorporated to the target prediction algorithm. The conveniently downloadable output in excel format is a clear advantage of this program.

Methods for Integrating miRNA and Gene Expression Data

In order to maximally assess the effect of miRNAs in a biological context it is essential to develop software, which take into account expression patterns and interactions of both miRNAs as well as mRNAs. To cohesively integrate miRNAs into meaningful biological networks, the designed software should be capable of taking into account quantitative data of miRNAs and mRNAs corresponding to dynamic changes within the context of a given experiment. To assess the effects of a relationship comprehensively It is desirable that quantitative dynamic changes in miRNAs as well as mRNAs are done in genome scale. The current miRNA target site prediction tools offer the opportunity of assessing the effects of a given individual miRNA or a set of miRNAs on transcripts in a static scale. There exists a challenge in developing tools that are capable of cohesively integrating genome scale changes of miRNAs and transcripts on a real-time quantitative scale of a given experiment so that the net effect of a given set of miRNAs on numerous transcripts can be viewed with respect to functional consequences of associations and their ultimate biological significance. As of today, the miRNA transcript associations are largely based on the premise that miRNAs target the 3'UTRs of mRNAs and thereby repress gene expression. Different criteria such as target site accessibility, free energy values, degree of complementarity and number of target sites are taken into account in deciding the most likely candidates that could possibly exert effects on a given transcript. It should be noted however that these predictions are done on an *in silico* basis on a non-quantitative basis. For instance, if we take in to consideration a time course experiment with different treatments, with existing technologies it is feasible to carry out genome scale profiling experiments to assess global quantitative representation of both miRNAs as well as mRNAs. In order to ascertain the net effect of miRNA-mRNA network associations to a final phenotype it would be necessary to qualify those miRNA and mRNA candidates which bear relationships to one another, thereafter sketch a network association based on dynamic changes which take place during the course of the experiments at different time points which do not take place by random chance alone. Upon finalizing the results of such an effort it would be feasible to depict genome scale anti-correlated miRNA-mRNA associations at different time points and the net biological effects of the observed associations. An advantage of such a method would be identifying particular nodes in a chaotic network, which perhaps dominantly contribute to a biological effect. In terms of application it would be most desirable to manipulate the candidates involved in those nodes, which stand apart from the rest. With efforts in designing software tools to fulfill assessment of large-scale miRNA-mRNA associations in a dynamic fashion and to correlate these associations to significant biological processes a number of attempts have been made. In the following passage we describe one such tool, which can be conveniently utilized to assess genome scale miRNA-mRNA associations in quantitative terms and biological significance of a given experiment.

Sigterms (http://sigterms.sourceforge.net) [49]

Sigterms has been built to determine miRNA-mRNA associations, utilizing macros supported by Microsoft Excel. The user interface consist of an excel workbook where a user can paste an entire set of genes or miRNAs of a given experiment generated through a standard profiling platform according to conventional annotations. Once the entire set of miRNAs or genes are input to the worksheet Sigterms scans the user defined set of miRNAs or genes against its in built annotated workbook of miRNA-mRNA associations derived for a given species from three commonly used miRNA target predication software tools (Target Scan, PicTar and miRanda) as well as standard gene ontology (GO) software. Once scanning is complete Sigterms generates an excel workbook with all miRNA-mRNA associations for the set of genes defined by the user for a given experiment along with an enrichment statistic which qualifies those associations which do not occur by mere chance. The output result which is based on differentially expressed candidates in the context of an experiment would thus more succinctly deliver results so much so anti-correlated pairs of miRNA-mRNA would be prominently highlighted upon sorting based on the enrichment statistic. A set of filters that are optional would afford the user to retrieve a particular set of associations, which comply additional user defined criteria. This feature allows the users to relate a particular miRNA or a gene of choice to a particular phenotype or biological significance in experimental perspective.

DISCUSSION

Through their work Gu and coworkers have shown that miRNA profiles of embryonic stem cells are dynamically regulated throughout the differentiation process [16]. The candidates which exhibit differential expression in ES and differentiated cells are likely to play key roles in maintenance of ES cell pluripotency, self renewal and lineage commitment [17, 50]. Although the role of key transcription factors regulating ES pluripotency and self renewal have been well studied in the past until recent times, their role in regulating miRNAs in ES cells were not all that well understood.

In 2008 Marson and colleagues through extensive genome wide ChIP-seq experiments described the pattern in which key stem cell regulators Oct4/Sox2 Nanog and Tcf3 along with polycomb group proteins exert their effects on miRNAs to regulate expression [38]. Sparse annotation of vertebrate miRNA promoters were limiting to cohesively relating transcription factor association with miRNAs [19]. However Barski *et al.,* [51] and Guenther *et al.,* [52] through independent studies in 2007 have found the occurrence H3K4me3 marks at transcriptional start sites for most genes in mouse and human. Using genomic coordinates derived from several cell types for the H3K4me3 chromatin signature Marson and coworkers were able to define high confidence promoters for over 80% of miRNA genes in mouse and human [38] They were then able to systematically carry out genome wide high resolution Chip seq experiments using antibodies for key ES cell transcription transcription factors Oct4, Sox2 NANOG, Tcf3 along with polycomb group repressor protein Suz12 and integrate miRNA genes to existing embryonic stem cell transcriptional regulatory circuitry by mapping efforts. Through these experiments it was possible to assert the existence of complex multilevel regulatory networks controlling ES cell identity. It is now understood that Oct4/Sox2/NANOG/Tcf3 occupied miRNAS are generally, preferentially expressed in ES cells. The fact that this observation can be extended to induced pluripotent stem cells, all the more emphasizes the significance of this association to ES cell identity and that ES cell specific transcription factors regulate expression of the subset of miRNAs associated with pluripotent stem cells [38]. It is striking that tissue specific miRNAs which are suppressed in ES cells are co occupied by polycomb repressor protein Suz12 along with ES specific transcription factors [38]. It is also evident that co-occupation of PcG proteins along with key ES transcription factors maintain lineage specific miRNAs repressed in ES state. Nevertheless they are maintained in a poised state for expression upon differentiation to specific lineages [38]. The loss of the suppressive chromatin mark H3K27me3 deposited by PcGs when differentiation is initiated and progressed, compliments the observation that the subset of miRNAs co occupied by polycomb group proteins along with key ES transcription factors are freed by the PcGs as and when necessary, thus retaining the tissue specific miRNA promoters under surveillance in ES cells in a state ready for transcription on demand [3, 38].

miRNA contribution to ES cells identity is characterized by examples such as murine mir290- mir-295 cluster with several matching seed sequences having similarities to those of mir-302 and mir-17-92 cluster predominant in human embryonic stem cells possessing promoter occupation by Oct4, Sox2, Nanog and Tcf3 [38] miRs in these clusters have been associated with cell proliferation [53, 54]. Additionally mir-430 zebra fish homolog of the mir 290-295 family has been shown to be necessary for rapid degradation of maternal transcripts in the developing zygote, besides its role of functioning as an element which precisely tunes the levels of nodal agonists lefty1 and 2 in relation to Nodal protein whose role in embryonic development is well characterized [55]. The mir290-295 cluster is also known to up regulate DNA methyl transferases 3a and 3b (DNMT3a and DNMT3b) through dampening the DNA methyl transcriptional repressor Rbl2 facilitating precise silencing of genes which require to be repressed in ES state through DNA methylation [38].

The ES key transcriptional factor, miRNA protein circuitry which operates downstream of Oct4 is perhaps more interesting. For example the negative feedback loop which exists between Let-7g and Lin28 is explained well by the fact that both Lin28 and pri-Let-7g are occupied by the ES key transcription factors [38, 57]. An incoherent feed forward circuit is thus observable in ES cells where Lin28 blocks processing, pri-Let-7g transcripts to mature Let-7g [38, 57]. However when transitioning from ES to differentiation mature Let-7g miRNA sequences dominate by antagonizing Lin28 to allow processing of pri-Let-7g to mature Let-7g necessary for regulating cellular differentiation [38]. In conclusion it can be stated that multilayered incoherent and coherent circuits are well witnessed in ES cell core transcriptional circuitry combined with PcGs and miRNAs [38]. The existence of these circuits are fundamentally important to maintain stem cell self renewal, pluripotence and transitioning to differentiation.

Genome wide high resolution ChipSeq data of different cell types have provided opportunity to extend our knowledge of stem cell regulation by key transcription factors and chromatin modifiers to miRNAs. Integrating these three layers of regulatory elements provides valuable insight into fundamental understanding of how stem cell pluripotence and self renewal are dynamically regulated in a subtly balanced, perfectly orchestrated fashion in stem cells. Besides the fact that stem cells hold great promise as future agents in cell mediated therapies, understanding the intricate networks and their dynamic associations regulating salient stem cell features would provide opportunity to not only safely use whole cell based therapies but also correct defective conditions in diseased cell types by driving molecular networks through reprogramming to match the nuances and molecular motifs seen in their most fundamental intrinsic state. Engineering and devising mechanisms to bring such challenging phenomena to reality requires in depth critical knowledge of fundamental molecular networks. As discussed in this chapter, numerous recently discovered *state of the art* techniques combined with powerful computing tools provide unprecedented opportunity to investigate biology governing stem cell phenomena to realistically drive experimentation closer to the next milieu of molecular therapeutics.

CONFLICT OF INTEREST

None declared.

ACKNOWLEDGEMENT

We would like to acknowledge that P.H.G. and J.B.T. are supported in part by a 1P01 GM081627-01 grant.

REFERENCES

[1]   Ralston A, Rossant J. Cdx2 acts downstream of cell polarization to cell-autonomously promote trophectoderm fate in the early mouse embryo.DevBiol 2008;313 (2):614-29.

[2]   Boyer LA, Lee TI, Cole MF, *et al*. Core transcriptional regulatory circuitry in human embryonic stem cells. Cell 2005;122 (6).

[3]   Boyer LA, Plath K, Zeitlinger J, *et al*. Polycomb complexes repress developmental regulators in murine embryonic stem cells. Nature2006;441 (7091):349-53.

[4]   Thomson JA, Itskovitz-Eldor J, Shapiro SS, *et al*. Embryonic stem cell lines derived from human blastocysts. Science 1998;282 (5391):1145-7.

[5]   Xu RH, Chen X, Li DS, *et al*. BMP4 initiates human embryonic stem cell differentiation to trophoblast. Nat Biotechnol 2002;20 (12):1261-4.

[6]   Takahashi K, Yamanaka S.Induction of pluripotent stem cells from mouse embryonic and adult fibroblast cultures by defined factors. Cell 2006;126 (4):663-76.

[7]   Takahashi K, Tanabe K, Ohnuki M, Narita M, Ichisaka T, Tomoda K, Yamanaka S. Induction of pluripotent stem cells from adult human fibroblasts by defined factors. Cell 2007; 131:861–872.

[8]   Yu J, Vodyanik MA, Smuga-Otto K, Antosiewicz-Bourget J, Frane JL, Tian S, Nie J, Jonsdottir GA, Ruotti V, Stewart R, Slukvin II, Thomson JA. Induced pluripotent stem cell lines derived from human somatic cells. Science 2007;318:1917–1920

[9]   Nakagawa M, Koyanagi M, Tanabe K, *et al*. Generation of induced pluripotent stem cells without Myc from mouse and human fibroblasts. Nat Biotechnol2008;26 (1):101-6.

[10]   Hanna J, Markoulaki S, Schorderet P, *et al*.Direct reprogramming of terminally differentiated mature B lymphocytes to pluripotency. Cell 2008;133 (2):250-64.

[11]   Aoi T, Yae K, Nakagawa M, *et al*. Generation of pluripotent stem cells from adult mouse liver and stomach cells. Science 2008;321 (5889):699-702.

[12]   Okita K, Yamanaka S.Induction of pluripotency by defined factors.Exp Cell Res 2010; [Epub ahead of print]

[13]   Judson RL, Babiarz JE, Venere M, BlellochR.Embryonic stem cell-specific microRNAs promote induced pluripotency.NatBiotechnol 2009;27 (5):459-61.

[14]   GunaratnePH.Embryonic stem cell microRNAs: defining factors in induced pluripotent (iPS) and cancer (CSC) stem cells?Curr Stem Cell Res Ther 2009;4 (3):168-77.

[15]   Lee TI, Jenner RG, Boyer LA, *et al*. Control of developmental regulators by Polycomb in human embryonic stem cells. Cell 2006;125 (2):301-13.

[16]   Gu P, Reid JG, Gao X, *et al*. Novel microRNA candidates and miRNA-mRNA pairs in embryonic stem (ES) cells.PLoS One 2008;3 (7):e2548.

[17]   Kanellopoulou C, Muljo SA, Kung AL, *et al*. Dicer-deficient mouse embryonic stem cells are defective in differentiation and centromericsilencing.GenesDev 2005;19 (4):489-501.

[18]   AmbrosV.The functions of animal microRNAs.Nature 2004;431 (7006):350-5.

[19]   Bartel DP. MicroRNAs: target recognition and regulatory functions. Cell 2009;136 (2):215-33.

[20]   Pepper AS, McCane JE, Kemper K, *et al*. The C. elegansheterochronic gene lin-46 affects developmental timing at two larval stages and encodes a relative of the scaffolding protein gephyrin.Development 2004;131 (9):2049-59.

[21]   Calabrese JM, Seila AC, Yeo GW, Sharp PA.RNA sequence analysis defines Dicer's role in mouse embryonic stem cells.ProcNatlAcadSci U S A 2007;104 (46):18097-102.

[22]   Morin RD, O'Connor MD, Griffith M, *et al.* Application of massively parallel sequencing to microRNA profiling and discovery in human embryonic stem cells. Genome Res 2008;18 (4):610-21.

[23]   Lai EC, Tomancak P, Williams RW, Rubin GM.Computational identification of Drosophila microRNA genes. Genome Biol 2003;4 (7):R42.

[24]   Lim LP, Glasner ME, Yekta S, Burge CB, BartelDP.Vertebrate microRNA genes.Science 2003;299 (5612):1540.

[25]   Boffelli D, McAuliffe J, Ovcharenko D, et al., Phylogenetic shadowing of primate sequences to find functional regions of the human genome. Science 2003;299 (5611):1391-4.

[26]   Pariset L, Chillemi G, Bongiorni S, Romano Spica V, ValentiniA.Microarrays and high-throughput transcriptomic analysis in species with incomplete availability of genomic sequences. N Biotechnol 2009;25 (5):272-9.

[27]   AffymetrixGeneChip, Accessed on June 22, 2010 at http://www.affymetrix.com

[28]   Agilent chip, Accessed on June 22, 2010 at http://www.agilent.com

[29]   Nimblegen Accessed June 22, 2010 at http://www.nimblegen.com/

[30]   CombiMatrix, Accessed on June 22, 2010 at http://www.combimatrix.com

[31]   Illumina bead array, Accessed on June 22, 2010 at http://www.illumina.com

[32]   Shingara J, Keiger K, Shelton J, *et al.* An optimized isolation and labeling platform for accurate microRNA expression profiling. RNA 2005;11 (9):1461-70.

[33]   Prosnyak MI, Veselovskaya SI, Myasnikov VA, *et al.* Substitution of 2-aminoadenine and 5-methylcytosine for adenine and cytosine in hybridization probes increases the sensitivity of DNA fingerprinting. Genomics 1994;21 (3):490-4.

[34]   Rampal, J.B., ed. In DNA Arrays: Methods and Protocols. Methods in Molecular Biology Humana Press. (2001).

[35]   Kumar R, Singh SK, Koshkin AA, Rajwanshi VK, Meldgaard M, Wengel J. The first analogues of LNA (locked nucleic acids): phosphorothioate-LNA and 2'-thio-LNA. Bioorg Med ChemLett 1998;8 (16):2219-22.

[36]   Arora A, Kaur H, Wengel J, Maiti S. Effect of locked nucleic acid (LNA) modification on hybridization kinetics of DNA duplex. Nucleic Acids SympSer (Oxf) 2008; (52):417-8.

[37]   Akamatsu W, DeVeale B, Okano H, Cooney AJ, van der KooyD.Suppression of Oct4 by germ cell nuclear factor restricts pluripotency and promotes neural stem cell development in the early neural lineage. J Neurosci 2009;29 (7):2113-24.

[38]   Marson A, Levine SS, Cole MF, *et al.* Connecting microRNA genes to the core transcriptional regulatory circuitry of embryonic stem cells.Cell 2008;134 (3):521-33.

[39]   Johnson DS, Mortazavi A, Myers RM, WoldB.Genome-wide mapping of *in vivo* protein-DNA interactions.Science 2007;316 (5830):1497-502.

[40]   Reid JG, Nagaraja AK, Lynn FC, *et al.* Mouse let-7 miRNA populations exhibit RNA editing that is constrained in the 5'-seed/cleavage/anchor regions and stabilize predicted mmu-let-7a:mRNA duplexes. Genome Res 2008;18 (10):1571-81.

[41]   Creighton CJ, Nagaraja AK, Hanash SM, Matzuk MM, Gunaratne PH.A bioinformatics tool for linking gene expression profiling results with public databases of microRNA target predictions. RNA 2008;14 (11):2290-6.

[42]   Berezikov E, Plasterk RH, Cuppen E.GENOTRACE: cDNA-based local GENOme assembly from TRACE archives. Bioinformatics 2002;18 (10):1396-7.

[43]   Krek A, Grün D, Poy MN, *et al.* Combinatorial microRNA target predictions.Nat Genet 2005;37 (5):495-500.

[44]   Lewis BP, Burge CB, Bartel DP. Conserved seed pairing, often flanked by adenosines, indicates that thousands of human genes are microRNA targets. Cell 2005;120 (1):15-20.

[45]   John B, Enright AJ, Aravin A, Tuschl T, Sander C, Marks DS. Human MicroRNA targets. PLoSBiol 2004;2 (11):e363.

[46]   Betel D, Wilson M, Gabow A, Marks DS, Sander C.The microRNA.org resource: targets and expression. Nucleic Acids Res. 2008;36 (Database issue):D149-53.

[47]   The   PITA   algorithm,   Accessed   on   June22,   2010   at *http://genie.weizmann.ac.il/pubs/mir07/mir07_prediction.html*

[48]   Kertesz M, Iovino N, Unnerstall U, Gaul U, Segal E.The role of site accessibility in microRNA target recognition. Nat Genet 2007;39 (10):1278-84.

[49]   Creighton CJ, Reid JG, GunaratnePH.Expression profiling of microRNAs by deep sequencing.Brief Bioinform. 2009;10 (5):490-7.

[50]   Murchison EP, Partridge JF, Tam OH, Cheloufi S, Hannon GJ.Characterization of Dicer-deficient murine embryonic stem cells.ProcNatlAcadSci U S A 200;102 (34):12135-40.

[51]   Barski A, Cuddapah S, Cui K, Roh*et al*. High-resolution profiling of histone methylations in the human genome.Cell 2007;129 (4):823-37.

[52]   Guenther MG, Levine SS, Boyer LA, Jaenisch R, Young RA.A chromatin landmark and transcription initiation at most promoters in human cells.Cell 2007;130 (1):77-88.

[53]   Laurent LC, Chen J, Ulitsky I, *et al*. Comprehensive microRNA profiling reveals a unique human embryonic stem cell signature dominated by a single seed sequence. Stem Cells 2008;26 (6):1506-16.

[54]   O'Donnell KA, Wentzel EA, Zeller KI, Dang CV, MendellJT.c-Myc-regulated microRNAs modulate E2F1 expression.Nature 2005;435 (7043):839-43.

[55]   Giraldez AJ, Mishima Y, Rihel J, Grocock RJ, Van Dongen S, Inoue K, Enright AJ, Schier AF.Zebrafish MiR-430 promotes deadenylation and clearance of maternal mRNAs. Science 2006;312 (5770):75-9.

[56]   AlonU.Network motifs: theory and experimental approaches.Nat Rev Genet 2007;8 (6):450-61.

[57]   Viswanathan SR, Daley GQ, Gregory RI.Selective blockade of microRNA processing by Lin28.Science 2008;320 (5872):97-100.

CHAPTER 10

Role of Translationally Regulated Genes in Embryonic Stem Cell Differentiation: Integration of Transcriptome and Translational State Profiling

Qian Yi Lee[1], Winston Koh[1], Prabha Sampath[2] and Vivek Tanavde[1,2,*]

[1]*Bioinformatics Institute, Agency for Science Technology and Research (A*STAR), Singapore and* [2]*Institute of Medical Biology, A*STAR, Singapore*

Abstract: This chapter describes a method for genome-wide identification of translationally regulated genes during embryonic stem cell differentiation using integrated transcriptome and translation state profiling. Previous attempts at identification of translationally regulated genes have focused on measuring the fractionation of the mRNA molecules in the translated and untranslated fractions without considering the transcriptional status of these genes. In this method we describe how integrating information about transcriptional and translational status of genes enables in identification of translationally regulated genes with much greater accuracy. This approach developed for microarrays can also be used for gene expression measured by next generation sequencing.

Keywords: Translationally regulated genes, human embryonic stem cell, diffrentiation, integration, transcriptome, translational state profiling, all-trans retinoic acid (ATRA).

INTRODUCTION

The spatial-temporal control of protein expression and activation is vital to many important biological processes, such as early development, differentiation and proliferation [1]. Since the protein level in a cell is largely determined by the extent of translation of mRNA into proteins, as well as the rate of protein degradation, translational control is crucial in controlling protein activity in the cell. Translational control can be defined as the regulation of protein synthesis *via* the translation step. Thus, factors affecting the translation of mRNA, such as RNA stability and RNA sequestration, all contribute to translational control in a cell.

*Address correspondence to Vivek Tanavde: Bioinformatics Institute, 30 Biopolis Street, #07-01, Matrix, Singapore 138671; Tel: 65-64788383; Fax: 65-64789048; E-mail: vivek@bii.a-star.edu.sg

It has been illustrated that translational control ensures that cells adopt correct identities and positions in early development [2]. In particular, it has been shown that translational control, through a hierarchy of translational regulators, is required for proper protein control and expression during embryonic stem cell differentiation [3].

The translation of mRNA into proteins is initiated by ribosomes binding to mRNA templates. Multiple ribosomes can bind to a single mRNA template to form a polysome. Translational efficiency, or the rate of translation, is generally proportional to the number of ribosomes bound to a single mRNA template. As polysomes are much denser than free mRNA, highly translated polysomal mRNA from each sample can be separated from untranslated free mRNA by using a sucrose density gradient. Microarrays can then be used to obtain genome-wide mRNA expression profiles for both the polysomal translated fraction and the free untranslated fraction [3]. By comparing polysomal and free mRNA expression in differentially treated samples, it is then possible identify the translational states of various mRNA in the cell.

There are numerous papers outlining methods to identify genes under translational control. Qin *et al.,* [4] used the ratio of polysomal mRNA over free mRNA to differentiate translationally active and inactive genes. Bushell *et al.,* [5] goes a step further by using a ranking analysis to compare these ratios for Tumor Necrosis Factor related apoptosis inducing ligand (TRAIL) treated *versus* control samples, and to identify genes that are translationally controlled due to TRAIL-treatment. Qin *et al.,* [4] and one of the authors of this chapter [3] also used a similar method that makes use of the ratio of polysomal fraction over the sum of all fractions. However, none of these methods took into account the confounding effects of transcription.

It is important to note that the level of protein expressed is affected both by transcriptional and translational control, as a gene must first be transcribed as an mRNA before it can be translated into a protein. Any changes in transcription will be carried forward to affect translation. So, unless the change in translation is significantly different from the change in transcription, it cannot be said that the gene is under translational control. Dinkova *et al.,* [6] did account for the effects of transcription by short-listing genes with no significant change in transcription during treatment, so that only purely translationally controlled genes are selected for. However, this method will not be able to pick out genes that are under the effects of both transcriptional and translational control.

Since most biological systems and biochemical pathways contain more than one level of control, it is highly likely that many genes are under the coupled effects of both transcriptional and translational control. This chapter outlines a method that would identify such genes by integrating transcriptome and translational state profiling. It will also further discuss the application of this method in identifying translationally controlled genes during the differentiation of human embryonic stem cells (hESC) using all-trans retinoic acid (ATRA).

INTEGRATING TRANSCRIPTOME AND TRANSLATIONAL STATE PROFILING

Background

Translational control can be defined as the regulation of protein synthesis *via* the translation step. However, as in many biological processes, there is often more then one level of control in the regulation of protein expression and activation. Along with genes that are purely regulated by transcriptional or translational control, there are many genes are regulated by the coupled effects of both transcription and translation. This section will introduce the concept of integrating transcriptome and translational state profiling to identify such genes. Differentiating human embryonic stem cells will be used as a model system to further elaborate on the methodology.

Experimental Design

The process of using microarrays in finding genes with that are transcribed into mRNA and not translated into protein under the influence of a particular treatment begins with the design of experiments. The treatment is hypothesized to induce a significant change in gene expression which would lead to changes in mRNA transcripts that is subsequently quantified using microarrays.

Typical analysis of the microarray results usually involves finding the fold change in gene expression after treatment; which essentially is the postulated number of transcripts after treatment divided by the number obtained before treatment. To enable finding translationally controlled genes, further experimental and analysis techniques are employed. Application of these techniques on each sample allows for separation of the transcripts into two different fractions: one containing transcripts with multiple ribosomes associated with it (translated fraction) and the other with fewer ribosomes (untranslated fraction).

Model System

HES3 human embryonic stem cells (hESCs) were treated with activin and bone morphogenetic protein 4 (BMP4) to induce differentiation into the mesendodermal lineage. hESCs were treated with the differentiation media for three days before being harvested for microarray. Control hESCs that were not treated with differentiation media were grown in parallel with treated hESCs for three days before being harvested for microarray.

Preparation of Total and Fractionated mRNA for Microarray Analysis

In order to identify translationally controlled genes, gene expression profiling of total mRNA needs to be performed, to obtain a transcriptome profile. Similarly gene expression of the fractionated mRNA can be profiled using microarray, to obtain a translational state profile. Cells from activin and BMP4-treated and control cell culture are first lysed to release the cell content, and then centrifuged to extract total mRNA from the resultant supernatant. The total mRNA can then be fractionated using a sucrose density gradient to separate poorly translated (pool 1) and highly translated (pool 2) mRNA. Total, pool 1 and pool 2 mRNA can then be loaded onto microarrays. (Fig. (**1**)). Note that there should be at least 3 technical replicates for each of pool 1, pool 2 and total RNA sample from both treated and control cell cultures to account for possible random errors such as sample loading and variation between microarray chips.

Initial Classification of Translationally Regulated Genes

The amount of transcripts in the pooled total and its fractions are quantified using microarrays. Before proceeding with the mathematical aspects of data analysis, the following simplifying assumptions are made. For each particular gene, given that the gene is not under the influence of translational control, the number of transcripts in each portion should obey:

$$\frac{Treated\ Total}{Control\ Total} = \frac{Treated\ Translated}{Control\ Translated} = \frac{Treated\ Un-Tranlsated}{Control\ Un-Translated}$$

This assumption holds true when considering the proportion of translated and untranslated transcripts. Given that that there is no translational control, the proportion of translated and untranslated transcripts should remain the same before and after treatment, hence the ratio of the total number of transcripts after and before treatment will be the same in the total pooled as well as the translated and un-translated fractions.

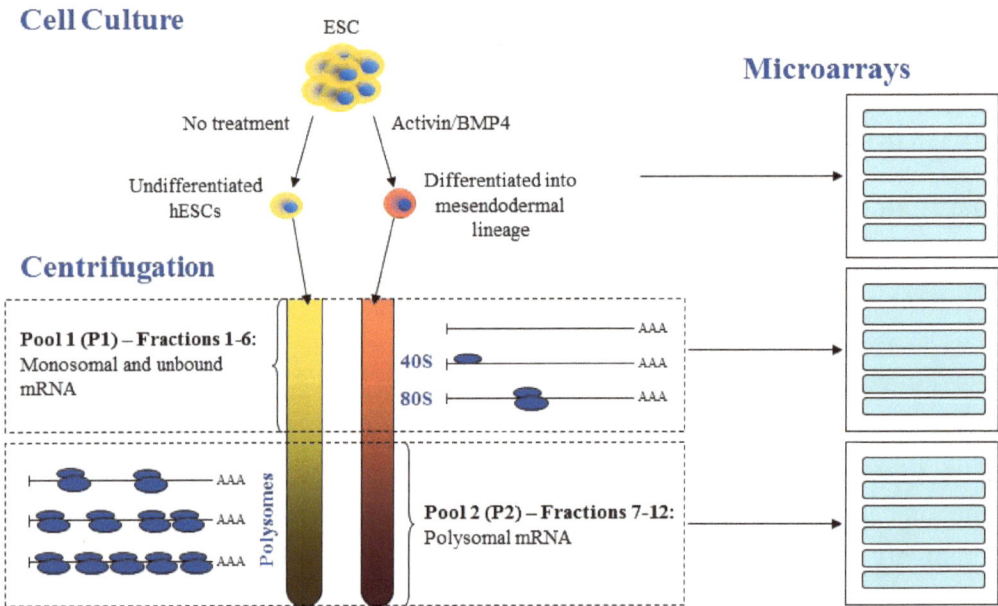

Figure 1: Experimental Design hESCs were treated with ATRA to induce differentiation into neural progenitor lineage. Centrifugation: Cellular lysates from control (untreated) and treated hESCs then fractionated to separate poorly translated (pool 1) and highly translated (pool 2) mRNA for translational state profiling. Microarray: Unfractionated cellular lysate (total RNA), pool 1 and pool 2 samples are split into 3 replicates each and loaded onto microarrays in a randomized manner.

However, when a particular gene is under translational control, the proportion of transcripts in the translated and un-translated fractions would change resulting in a shift in the number of transcripts towards the translated or untranslated fraction. Respectively, this corresponds to an increase or a decrease in translational efficiency. In terms of inequalities, a shift towards a greater number of translated transcripts is represented as:

$$\frac{Treated\,Translated}{Control\,Translated} > \frac{Treated\,Total}{Control\,Total} > \frac{Treated\,Un-Tranlsated}{Control\,Un-Translated}$$

Similarly, when there is an increase in the proportion of the number of transcripts in the untranslated fraction, the inequality can be re-expressed as:

$$\frac{Treated\,Translated}{Control\,Translated} < \frac{Treated\,Total}{Control\,Total} < \frac{Treated\,Un-Tranlsated}{Control\,Un-Translated}$$

With these inequalities in mind, a filter is implemented to find genes that satisfy the inequalities and are assumed to under translational control. The ratios of the transcripts are translated into fold changes. Subsequently, a simple measure to quantify the shift in proportion towards translation or untranslated fraction is given by the difference between the fold changes. For example, the shift towards translation is given by: Fold change of the total pool (after and before treatment) – Fold change of the translated fraction (after and before treatment).

In order to identify genes that are under both transcriptional *and* translational control, a filter for false discovery rate (FDR) ≤ 0.05 and fold-change ≥ 2 is first applied to pick out transcriptionally regulated genes. The shift towards the translated or untranslated fraction is then determined by the difference between fold-changes as described above.

Using such measures, each gene can now be categorized into 4 main classes (Fig. (**2**)):

1. Genes that are up-regulated and with corresponding increase in translational efficiency.

2. Genes that are up-regulated but with a decrease in translational efficiency.

3. Genes that are down-regulated but with an increase in translational efficiency.

4.   Genes that are down-regulated and with a corresponding decrease in translational efficiency.

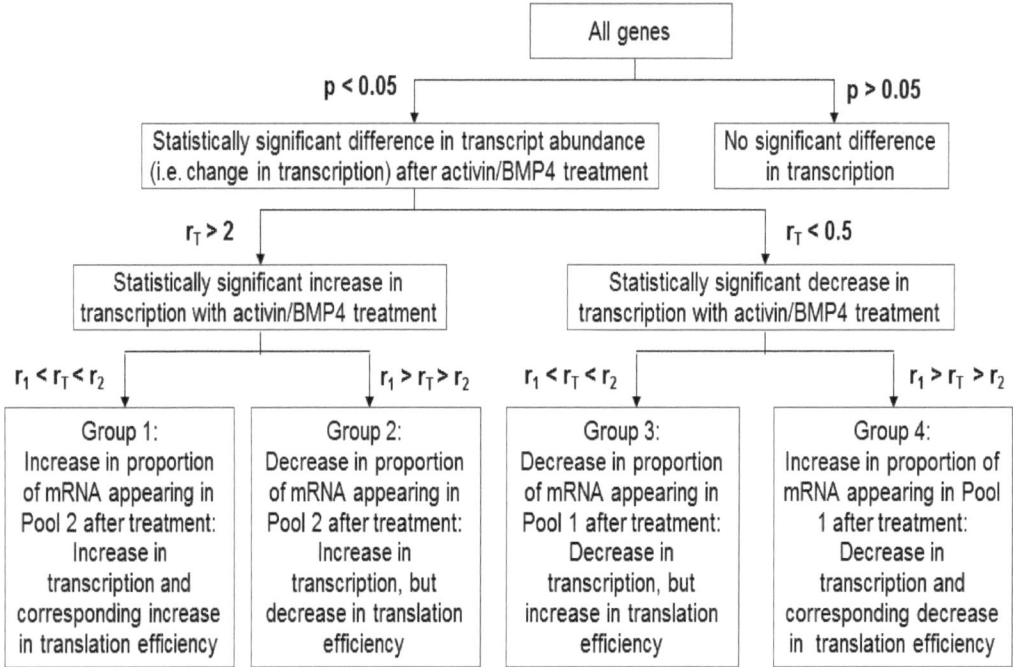

Figure 2: Flow chart for categorizing genes into four main classes.
Genes can be categorized into four main classes:
1.   Genes that are up-regulated and with corresponding increase in translational efficiency.
2.   Genes that are up-regulated but with a decrease in translational efficiency.
3.   Genes that are down-regulated but with an increase in translational efficiency.
4.   Genes that are down-regulated and with a corresponding decrease in translational efficiency.

Transcriptionally regulated genes are first identified by looking at the difference in total mRNA expression. A p-value cut-off of 0.05 ($p < 0.05$, where r_T = total mRNA in treated samples/total mRNA in untreated samples) was used to short-list genes with a significant difference in transcript abundance in the total mRNA level. A fold-change cut-off of 2 ($r_T >= 2$, where r_T = total mRNA in treated samples/total mRNA in untreated samples) is then used to identify genes that are significantly up-regulated while a cut-off of 0.5 ($r_T <= 0.5$) is used to identify genes that are significantly down-regulated.

Significantly up- or down-regulated genes are then further separated according to changes in translational efficiency. Genes with $r_1 < r_T < r_2$ (where r_1 = Pool 1 mRNA in treated samples/Pool 1 mRNA in untreated samples, and r_2 = Pool 2 mRNA in treated samples/Pool 2 mRNA in untreated samples) are considered to have experienced an increase in translational efficiency, as there is a shift of mRNA to the polysomal, translated pool. Conversely, genes with $r_1 > r_T > r_2$ are considered to have experienced a decrease in translational efficiency, as there is a shift of mRNA to the free, untranslated pool.

For the activin-BMP4 model system, 443 genes that are potentially under translational control were identified Fig. (**3**).

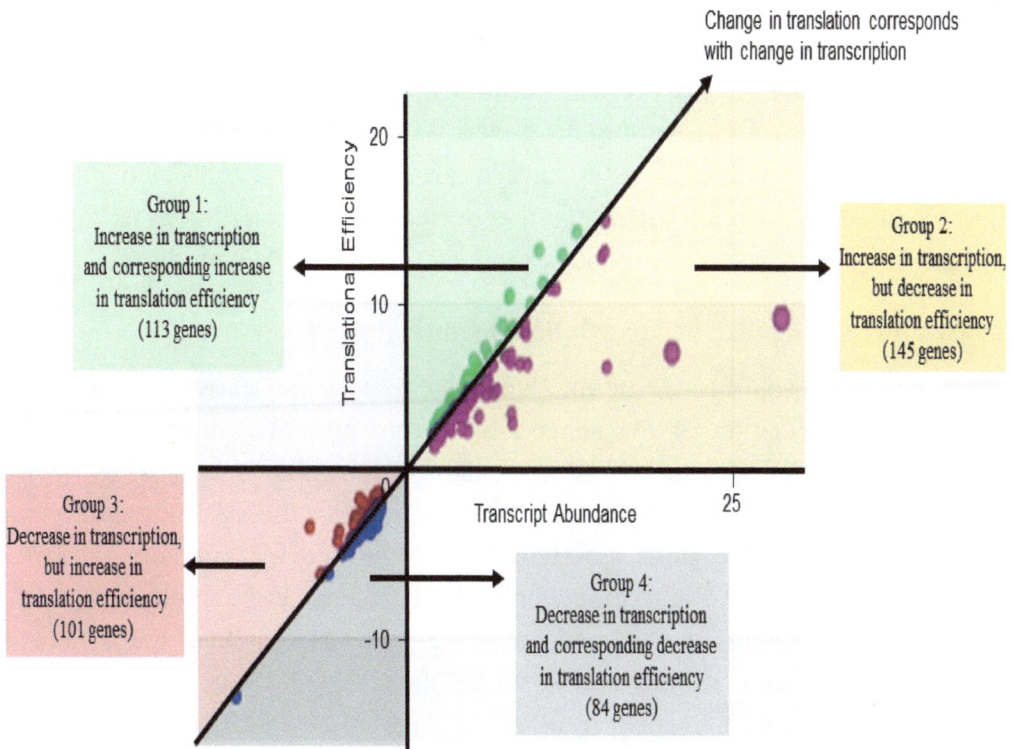

Figure 3: Translationally regulated genes identified with microarray.

Scatter plot of Translational Efficiency (Pool 2 treated/Pool 2 untreated) against Transcript Abundance (Total Treated/Total Untreated). Changes in untranslated mRNA (Pool 1 treated/Pool 1 untreated) are represented by the size of each scatter point. Points along the x = y line shows genes that are purely transcriptionally controlled. That is, the change is translation is equal to the change in transcription, so there is no translational control of that particular gene in this system. This plot gives a graphical interpretation of the four main categories formed in Fig. (**2**). Group 1 contains genes that are up-regulated and with corresponding increase in translational efficiency. These genes have a larger change in translated mRNA compared to total mRNA, such that the increase in transcription is further amplified during translation to produce exponentially more protein products. Group 2 contains genes that are up-regulated but with a decrease in translational efficiency. In this case, genes are up-transcribed but translationally repressed, such that the amount of protein expressed is less than expected for the amount of mRNA being transcribed. Group 3 contains genes that are down-regulated but with an increase in translational efficiency. Contrary to Group 2, Group 3 contains genes that are down-transcribed but still being translated. Group 4 contains genes that are down-regulated and with a corresponding decrease in translational efficiency. These genes have a larger change in translated mRNA compared to total mRNA, such that the decrease in transcription is further amplified during translation to produce exponentially less protein products.

Determining Significance of Translational Control

In determining the significance of the gene in its respective classes, histograms describing the distribution of the shifts towards the untranslated and translated fractions are plotted along with the cumulative probability graphs (Fig. (**4**)). Observations of the cumulative probability plots suggest that the distribution of both translated and untranslated shifts closely approximates to a Gaussian distribution. Using the cumulative probability plots, thresholding filters can be implemented on the probability to obtain list of genes deemed to exhibit significant translational shifts. Based on the scale of experimental studies, this threshold point can be stricter to reduce the number of candidate genes for validation.

In the activin-BMP4 system, the cumulative probability plot with a cut-off of 0.2 was used, and 126 candidate genes were short-listed. The significance cut-off of 0.2 was chosen to obtain a reasonable amount of candidate genes (~100 genes) for further analysis. Genes that make these criteria in each of the categories can then be interpreted to be expected by chance with a probability of less than 0.2.

A

B

C

D

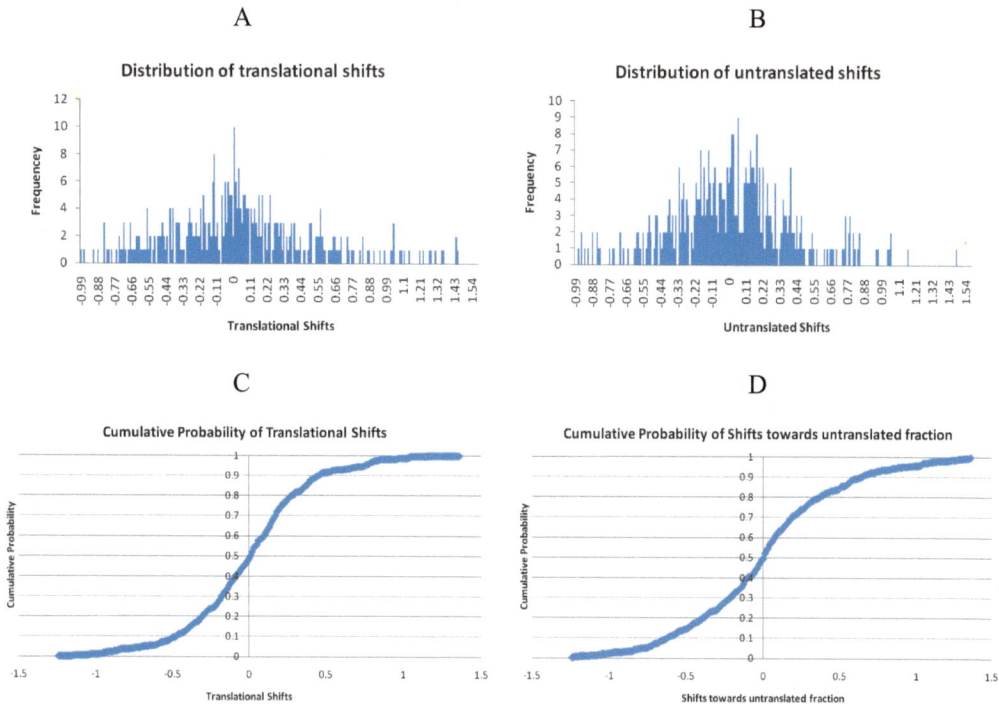

Figure 4: Histograms and cumulative probability graphs of translational shifts. The shift in proportion of mRNA towards the translated or untranslated fraction can be measured by the difference between the fold changes. **A** shows the distribution of shifts for translated pool compared to total RNA (total fold-change – Pool 2 fold-change). **B** shows the distribution of shifts for untranslated pool compared to total (total fold-change – Pool 1 fold-change). From both **A** and **B**, it can be observed that these shifts of mRNA towards the translated or untranslated fractions follow a Gaussian distribution. Thus, using cumulative probability plots of the two distributions (**C** and **D**), thresholding filters can be implemented on the probability to short-list genes deemed to exhibit significant translational shifts.

Of the short-listed genes, more than half were up-transcribed but translationally repressed. Of these genes, there was an overrepresented group of six genes that are associated with the keratin family of genes, including keratin, keratin-associated proteins, and keratin pseudogenes. This finding corresponded to the results published by Lu *et al.,* [7] that keratins are transcribed but not translated in differentiating embryonic stem cells.

Identification of Non-Coding RNA

In addition, a large proportion of these short-listed genes are annotated as non-coding RNA on the NCBI database (http://www.ncbi.nlm.nih.gov). As non-coding RNA are transcribed but not translated, they will not associate with ribosomes and will appear largely in the untranslated pool 1 mRNA. As such, they largely appear as genes with a decrease in translational efficiency (Group 2 or 3). Thus, in addition to identifying translationally controlled genes, this method can also be used to pick out non-coding RNAs that are not being translated.

Summary

In summary, a typical descriptive statistical analysis of a study with the imposed initial filter for genes that fit the inequalities and the four main classes would reduce the entire list of genes from 37805 to approximately 400 genes with about the same magnitude of number of genes in each class of which yields an estimated number of approximately 30 candidate genes in each class meets the level of significance at 0.2. Non-coding RNA can also be identified from Group 2 or Group 3 genes.

ALTERNATE APPROACHES TO IDENTIFYING TRANSLATIONALLY CONTROLLED GENES

Deep Sequencing (RNA-seq)

A combination of polysome profiles generated by sucrose gradient centrifugation methods followed by deep sequencing can also be done to identify and quantify precisely the differentially translated genes. The advantage of deep sequencing is that it allows the detection of various isoforms, allowing one to determine which isoform (s) are being translationally controlled. In addition, it allows the quantification of genes with low levels of expression.

However, one of the common problems encountered in sequencing is the presence of reads that map to multiple regions in the reference genome. While genic regions are usually well conserved and contain less repetitive regions than inter-genic regions, families of genes often contain domains with high degrees of sequence similarity (*e.g.,* HOX cluster genes). In addition, pseudogenes and computationally predicted genes also share large regions of similarity with their genic counterpart (*e.g.,* KRT16 shares more than 90% nucleotide sequence similarity with KRT16P3). Thus, when conducting RNA-seq, it is advisable to use as long a read length as possible to reduce the likelihood of a read mapping to multiple regions.

Ribosome Profiling

Rapid advances in deep sequencing technology also add another dimension in the field of translational control through a technique known as ribosome profiling. Using deep sequencing of short RNA fragments, it is now possible to determine the exact ribosomal positions on mRNA. After stabilization of polysomes by adding cycloheximide, ribosome "footprints" can be generated by nuclease digestion of polysomal RNA. Free mRNA between ribosomes is degraded by the nuclease while short mRNA fragments (approximately 30 nucleotides in length) that are bound to ribosomes are 'shielded' from degradation. These ribosome "footprints" can then be sequenced using any small mRNA sequencing platforms available (*e.g.*, Illumina HiSeq) and subsequently aligned to a reference genome to determine the exact location each ribosome was bound. The abundance of these ribosome "footprints" on each mRNA can be used to determine the rate at which the mRNA is being translated. More importantly, ribosome profiling adds another dimension to translational control by revealing differences in ribosomal density along the length of a gene. This would reflect any differences in translation initiation and the rate of elongation of a polypeptide chain that would not have been detected by microarray experiments.

Ingolia *et al.,* [8] had demonstrated that these ribosome "footprints" can be mapped with a high degree of precision to reveal a three-base periodicity that corresponds to the codons within protein-coding sequences across the transcriptome. However, there are some factors that should be taken into consideration when investigating translational control using ribosome profiling. Firstly, these ribosome "footprints" are, by nature, short reads. The presence of highly similar stretches of sequences within gene families and between pseudogenes and their genic counterparts make it problematic to assign the short ribosome "footprint" to a specific region as it would most likely result in multiple mappings for each read. Secondly, ribosome "footprints" only allow the characterization of translated mRNA, since only mRNA that are bound to ribosomes will have their fragments sequenced. Even if total mRNA were to be sequenced simultaneously, ribosome profiling does not allow the direct quantification of genes that are translationally repressed. Thus, this may decrease the accuracy of investigating genes that undergo transcriptional control coupled with translational control. Finally, ribosome profiling does not allow the quantification of untranslated non-coding mRNA, as these mRNA do not associate with ribosomes to form polysomal complexes. In this respect, it falls behind microarrays in the kind of information that can be provided.

CONCLUSION

As next generation sequencing techniques proceed to mature, the degree of resolution and insight into many intricate aspects of transcription and the associations with ribosomal complexes would introduce new directions for studies. However, as demonstrated, the ease of use of microarrays allows for a very rapid initial screen for potential candidate genes to be studied which are applicable to a wide range of studies and treatments. All of which is required are additional analytical methods to perceive the data which were described above.

CONFLICT OF INTEREST

None declared.

ACKNOWLEDGEMENT

This work was funded by Agency for Science Technology and Research (A*STAR), Singapore.

REFERENCES

[1]   Kuersten S, Goodwin EB. The power of the 3' UTR: translational control and development. Nat Rev Genet 2003; 4 (8): 626-637.
[2]   Richter JD, Theurkauf WE. The message is in the translation. Science 2001; 293 (5527):60-62.
[3]   Sampath P, Pritchard DK, Pabon L, *et al.* A hierarchical network controls protein translation during murine embryonic stem cell self-renewal and differentiation. Cell Stem Cell 2008; 2 (5):448-460.
[4]   Qin X, Ahn S, Speed TP, Rubin GM. Global analyses of mRNA translational control during early Drosophila embryogenesis. Genome Biol 2007; 8 (4):R63.
[5]   Bushell M, Stoneley M, Kong YW, *et al.* Polypyrimidine tract binding protein regulates IRES-mediated gene expression during apoptosis. Mol Cell 2006; 23 (3):401-412.
[6]   Dinkova TD, Keiper BD, Korneeva NL, Aamodt EJ, Rhoads RE. Translation of a small subset of *Caenorhabditis elegans* mRNAs is dependent on a specific eukaryotic translation initiation factor 4E isoform. Mol Cell Biol 2005; 25 (1):100-113.
[7]   Lu H, Hesse M, Peters B, Magin TM. Type II keratins precede type I keratins during early embryonic development. Eur J Cell Biol 2005; 84 (8):709-718.
[8]   Ingolia NT, Ghaemmaghami S, Newman JR, Weissman JS. Genome-wide analysis *in vivo* of translation with nucleotide resolution using ribosome profiling. Science 2009; 324 (5924):218-223.

CHAPTER 11

Paired SAGE-microarray Expression Data Sets Reveal Antisense Transcripts Differentially Expressed in Embryonic Stem Cell Differentiation

Reatha Sandie[1], Christopher J. Porter[1], Gareth A. Palidwor[1], Feodor Price[1], Paul M. Krzyzanowski[1], Enrique M. Muro[2], Sebastian Hoersch[3,4], Mandy Smith[1], Pearl A. Campbell[1], Carolina Perez-Iratxeta[1], Michael A. Rudnicki[1] and Miguel A. Andrade-Navarro[1,2,*]

[1]Ottawa Hospital Research Institute, 501 Smyth Road, Ottawa, ON, Canada K1H 8L6; [2]Max Delbrück Center for Molecular Medicine, Robert-Rössle-Strasse. 10, 13125 Berlin, Germany; [3]Informatics and Computing Core, Koch Institute for Integrative Cancer Research, Massachusetts Institute of Technology, 77 Massachusetts Avenue, Cambridge, MA 02139, USA and [4]Bioinformatics Group, Max Delbrück Center for Molecular Medicine, Robert-Rössle-Strasse. 10, 13125 Berlin, Germany

Abstract: Serial Analysis of Gene Expression (SAGE) is a sequence-based measure of gene expression that provides quantitative information on the population of transcripts through the generation and counting of specific sequence tags. Many SAGE datasets are publicly available for analysis, constituting a valuable resource for the study of gene expression. These datasets contain tags that are not obviously derived from known transcripts and thus hint at the existence of a large number of novel transcripts; however, the prioritization of candidates for further experimental verification is difficult. Here we demonstrate a method to identify non-coding antisense transcripts which may be implicated in stem cell differentiation by combining SAGE data with gene expression data derived by a complementary method. We produced SAGE libraries and paired microarray gene expression data pre- and post-differentiation of three mouse stem cell types (embryonic, mammary and neural). We found 1,674 SAGE tags antisense to 1,351 protein coding genes. A majority of these antisense tags overlap the 3'UTRs of sense genes; their abundance correlates with the expression of the corresponding sense genes and appears to be tissue specific. We did not find significant association between the expression of these tags and alternative splicing. We measured the expression of three genes expressed in the mouse embryo (Zfp42/Rex1, Ywhag/14-3-3g and Pspr1) and corresponding putative antisense transcripts by qPCR before and after differentiation of mESC. We conclude that it is possible to identify putative novel

*Address correspondence to Miguel A. Andrade-Navarro:** Computational Biology and Data Mining Group, Max Delbrück Center for Molecular Medicine, Robert-Rössle-Str. 10, 13125 Berlin, Germany; Tel: +49-30-9406-4250; Fax: +49-30-9406-4240; E-mail: Miguel.Andrade@mdc-berlin.de

Ming Zhan (Ed)

antisense transcripts with a potential role in ES cell differentiation by integrating data from existing SAGE libraries with expression data derived by a complementary method. All data used in this work are available from the Gene Expression Omnibus (GEO) and StemBase databases.

Keywords: Serial Analysis of Gene Expression, DNA microarray profiling of gene expression, Embryonic stem cells, Neural stem cells, Mammospheres, Stem cell differentiation, Antisense transcripts, Non-coding RNA, Alternative splicing, Expressed Sequence Tag libraries.

INTRODUCTION

Serial Analysis of Gene Expression (SAGE) is a high-throughput method for measuring gene expression, which estimates transcript abundance in a sample from counts of sequence-specific tags [1]. Many SAGE libraries are publicly available for analysis, for example in the Gene Expression Omnibus (GEO) [2]. The advent of novel multiplex DNA sequencing technologies (*e.g.* RNA-seq [3]) may make SAGE a less attractive option for future studies; however, existing SAGE libraries constitute a valuable resource for the research of gene expression.

The value of SAGE data is exemplified by the large number of novel transcripts for which it provides evidence, both protein-coding and non-coding. In particular, SAGE libraries from many eukaryotic organisms suggested an unexpected number of antisense transcripts (putative *cis*-acting ncRNAs) in plants (wheat [4], sugar cane [5], rice [6], *Arabidopsis thaliana* [7]), fungi (*Magnaporthe grisea* [8]), *Drosophila melanogaster* [9], *Plasmodium falciparum* [10], mitochondria from *Caenorhabditis elegans* [11] and mammals [12-18]. However, the number of such transcripts with verified biological function remains small and the function of antisense transcripts in general is unclear [19-21].

The correlation between expression of corresponding sense and antisense transcripts and the observation that antisense transcripts are expressed at a much lower level [14, 16, 20, 22-24] could be interpreted as indicating a lack of functionality. However, accumulating evidence suggests otherwise: precise spatial expression patterns of antisense transcripts have been described in mouse brain [25] and it has been demonstrated that the expression of antisense RNAs can alter the expression of corresponding sense RNAs [20] (recently described in great detail for gene PU.1 [26]).

The experimental verification of the large number of putative novel antisense transcripts suggested by SAGE libraries is however a colossal task. As it is not yet clear which of those candidates are functional, a prudent first step is to design strategies to identify SAGE tags more likely to represent biologically functional transcripts.

One possibility, explored here, is to pair SAGE data with gene expression data obtained using a different experimental platform from the same (or comparable) samples. Microarray-derived expression data are abundant and offer a view of gene expression complementary to SAGE, since the correspondence of the measurements (probeset hybridization) to genes is more accurate, albeit limited to the genes (mostly protein coding) for which probe sets have been placed in the microarray. The idea is to contrast the patterns of expression of antisense transcripts detected by SAGE and of the overlapping sense gene as measured by the microarray.

Table 1: Samples used in this work. These data are available in StemBase [31, 33, 53].

Description	SAGE			DNA microarray		
	Experiment	Sample	GEO	Experiment	Sample	GEO
R1 ESC	E221	S358	GSE3233	E165	S206	GSE2972
R1 Embryoid bodies	E222	S359	GSE3007	E165	S216	GSE2972
Mammospheres	E226	S355	GSE3841	E199	S255	GSE3777
Differentiated mammospheres	E227	S356	GSE3841	E199	S256	GSE3777
Neural Stem Cells	E224	S331	GSE3781	E206	S271	GSE3779
Committed Neural Progenitors	E225	S332	GSE3782	N/A	N/A	N/A

To apply this approach to the detection of antisense transcripts with potential roles in stem cell differentiation, we generated SAGE libraries from samples of three types of murine stem cells (embryonic, mammary and neural) and their differentiated derivatives and Affymetrix microarray gene expression data generated from equivalent samples (Table **1**). We identified SAGE tags antisense to genes (Table **2**) by comparing SAGE tag strand to that of known overlapping coding sequences. These antisense tags had tissue specific patterns of expression, were most frequently located in the 3'UTR of the sense gene and antisense expression tended to co-occur with that of the sense gene, properties that are consistent with a modulator role.

Table 2: Details of six SAGE libraries.

Description	Id	Antisense tag counts	Total tag counts	Fraction of antisense tags	Antisense tags with counts (normalized[1])	Differentially expressed antisense tags[2]
R1 ESC	S358	1713	99464	1.7%	450	32
R1 Embryoid bodies	S359	1524	78176	2.0%	488	45
Mammospheres	S355	2858	100670	2.8%	642	65
Differentiated mammospheres	S356	1010	42224	2.4%	489	48
Neural Stem Cells	S331	917	69910	1.3%	336	11
Committed Neural Progenitors	S332	1931	43026	4.5%	782	161

[1]Average from 100 random samplings of 42,000 tags.
[2]Number of tags with a count fold change of 3 or greater in stem cell *versus* differentiated library (*e.g.* R1 ESC *versus* R1 Embryoid bodies). A pseudocount of one is added to the denominator to avoid division by zero. Averaged from 100 random samplings of 42,000 tags.

To identify putative *cis*-acting non-coding RNAs with a biological role in stem cell differentiation, we selected tags with higher expression in stem cells than in the differentiated counterpart antisense to genes expressed in the stem cell, according to the microarray data. Time series RT-qPCR of predicted antisense transcripts for three of the tags (associated with sense genes Zfp42/Rex1, Ywhag/14-3-3g and Prps1) suggested that although each antisense transcript was co-expressed with the sense transcript, the profiles of sense and antisense expression in mESC differentiation were different. Our results support the notion that the integration of SAGE libraries with complementary gene expression data can be used to identify non-coding RNAs of potential biological significance.

METHODS

Compilation of SAGE Libraries

Samples for SAGE were gathered as part of the Stem Cell Genomics Project, performed in collaboration with researchers from the Canadian Stem Cell Network (http://www.stemcellnetwork.ca/).

Embryonic stem cell libraries were constructed from undifferentiated R1 mESC grown as previously described [27] over mitotically inactivated DR4 fibroblast feeders in the presence of Leukemic Inhibitory Factor (LIF). Differentiated R1s were generated by growth as unattached spheres for nine days in the absence of LIF and murine embryonic fibroblast feeder cells.

Putative neural stem cells (NSCs) were isolated from E12-E14 striatum of C57Bl/6 mice and were cultured as unattached neurospheres for 10-12 days in DMEM/F12 supplemented with bFGF, D-glucose, L-glutamine, penicillin/streptomycin, insulin and apotransferrin, progesterone, putrescine, selenium, Fungizone and heparin. Neural committed progenitors were then generated from the NSCs by plating on 5 µg/ml poly-L-ornithine and 10 µg/ml laminin and growth in neuronal differentiation media (DMEM-F12) in the presence of D-glucose, L-glutamine, penicillin/streptomycin, insulin and apotransferrin.

Murine mammospheres were derived from 9-12 weeks virgin female FVB mice and were allowed from for 3 passages in 3:1 DMEM:F12 media supplemented with B-27, bFGF, EGF and heparin. Differentiation of the mammosphere stem/progenitors to more committed myoephithelial/luminal cell types was achieved by growth over a collagen matrix in the presence of DMEM supplemented with 5% FBS.

Total RNA from each sample was isolated using the RNeasy Mini Kit (Qiagen). Following RNA isolation each underwent quality assessment on the BioAnalyzer 2100 (Agilent). To ensure the preparation of high quality SAGE libraries a 28S/18S rRNA ratio of 1.8 was used as a cutoff score for library construction.

Long SAGE library construction was performed from 10 micrograms of total RNA using the I-SAGE Kit and I-SAGE Ditag PCR Module (Invitrogen) according to manufacturer's recommended protocol. The average size of each concatemer library was approximately 1,000 bp for a total of 10 tags per concatemer. Cloned concatemers were sequenced on a 3730 Genome Analyzer (Applied Biosystems) at StemCore Laboratories (http://www.stemcore.ca)

Microarray Analysis

Gene expression values and calls were generated using Affymetrix MAS5 software (Affymetrix, 2001) with parameters of alpha1=0.04, alpha2=0.06 and tau=0.015.

These are standard default values for these parameters. The alpha terms indicate the p-value's threshold under which probe sets are called present: Present if p-value ≤ alpha1, Marginal if p-value > alpha1 and p-value ≤ alpha2 and Absent if p-value > alpha2). The tau term is a small constant used in the Wilcoxon Signed Rank Test of probe pairs. Expression values for replicates were averaged. For Present/Absent calls, probe sets were considered present if the majority (2/3) of replicates had that probeset called as Present, otherwise the probeset was considered Absent.

RT-qPCR

RNA was isolated using the RNEasy kits (Qiagen) and subjected to on-column DNase digestion as per the manufacturer's instructions. cDNA synthesis was performed using Superscript II reverse transcriptase with random hexamers (Invitrogen). SYBR Green real-time PCR reactions were performed in duplicate using an MX3000p PCR machine (Stratagene, La Jolla, CA) with fold-change normalized against GAPDH. Primers for real-time PCR were designed using the online Primer3 software (http://frodo.wi.mit.edu/primer3/) with gene specific information obtained from Ensembl. All primer sequence information is provided (Table **5**). Relative fold change in expression was calculated using the ΔΔCT method (CT values < 30) and primer specificity was validated by denaturation curve analysis (55–94 °C). Amplification-curve plotting and calculation of cycle threshold (Ct) values were performed using the MX3000p software (v3; Stratagene), with further calculations performed using Microsoft Excel. Each experiment was performed at least twice.

Western

Total protein was harvested in RIPA lysis buffer, subjected to SDS PAGE and electroblotted onto Immobilon-P membrane (Millipore). Membranes were probed with primary antibody and HRP-conjugated secondary antibody in blocking solution. Target proteins were visualized by ECL (Amersham-Pharmacia) with Biomax MAR film (Kodak). Primary antibodies used were: rabbit polyclonal anti-PRPS1 (Abcam ab38209). The secondary antibody was an HRP-conjugated anti-rabbit (Biorad).

Selection of Antisense Tags

We mapped all tags in our six libraries to the mouse genomic sequence (NCBI v36) and selected those that mapped to a single location (100% identity; no gaps).

We then compared tag positions to the boundaries of mouse transcripts from Ensembl (version 42, January 2007). A tag was identified as being from a putative antisense transcript if it overlapped a known mouse transcript; if a tag overlapped multiple sense transcripts, only the first returned by Ensembl was selected.

Comparison of Antisense Expression to Proteomics Data

A set of proteins expressed in R1 mESC was obtained from Tables 1, 2 and 3 from [28]. Only 157 of the 241 proteins in these tables are identified as being from the mouse; this is because the peptides obtained by mass spectrometry were identified by searching a mammalian protein database. We chose to work with these 157 proteins rather than risk selecting the incorrect mouse ortholog for the remaining 64.

Correlation Between Splicing Properties of the Sense Gene and Expression of the Antisense Transcript

We categorized genes into those with and without alternatively spliced forms by intersecting the UCSC Genome Browser tracks "AceView" [29] (for a complete collection of genes and their transcripts) and "Alt Events" [30] (for a conservative estimation of alternative splicing events). Genes with any transcripts overlapping an Alt Events entry were classified as alternatively spliced.

Using this approach and considering only AceView genes corresponding to well supported genes (*i.e.* the AceView name starting with an upper-case letter), we arrived at a total of 28,564 genes, of which 10,958 (38%) were classified as alternatively spliced and 17,606 (62%) were not.

RESULTS AND DISCUSSION

We produced six SAGE libraries from six samples grouped in three pairs of stem cells and differentiated counterparts (Table **1**). Affymetrix microarray gene expression data from five of the six samples, generated previously in the same laboratory [31], were available for comparison. Initial analysis of our data did not suggest a significant number of sequencing errors [32] or any detectable sequencing error bias (data not shown). Comparisons of sequence tags from known transcripts to the equivalent hybridization values from the microarray analyses indicated general agreement between the SAGE libraries and microarray experiments (Fig. **1**).

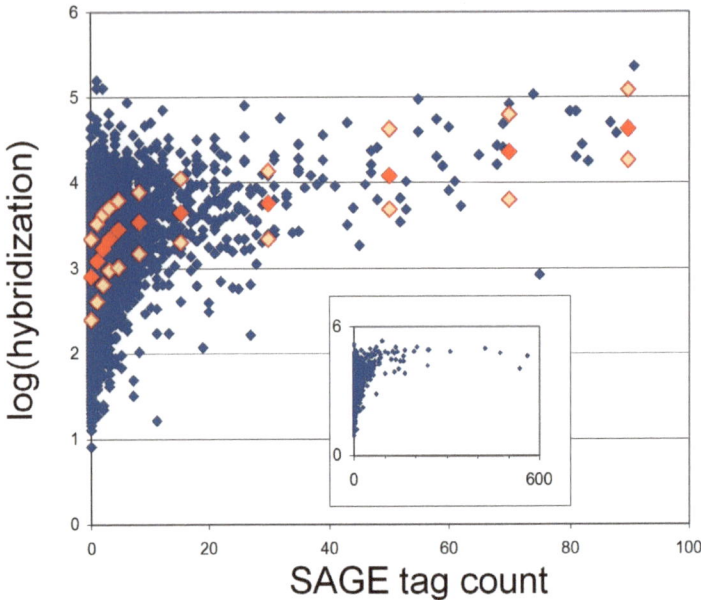

Figure 1: Comparison of expression measured by SAGE and microarray. Each blue point represents expression of a transcript measured by SAGE (tag counts, x axis) and microarray (log of MAS5 hybridization value, y axis) in R1 mESC. Red diamonds indicate the 20th, 50th and 80th percentiles of DNA hybridization at different SAGE tag counts. The inset shows the values for the full data range. SAGE tags were assigned to mouse genes by identifying the 17 bases adjacent to the 3'-most *Nla*III restriction site in the RefSeq sequence. Corresponding Affymetrix probe sets from the MOE430 expression microarray were identified in the NetAffx database [52]. Microarray and SAGE data can be retrieved from StemBase (samples 206 and 358, respectively). Similar graphs were obtained when comparing the other two pairs of DNA microarray and SAGE experiments (data not shown).

In some cases, tag sequences (21nt) from our SAGE libraries map to multiple genomic locations; this may represent gene duplications or pseudogenes. Similarly, some tags are not found at any contiguous genomic location, which may occur because of variation in the sequence (either sequencing errors or natural variation) or if the SAGE tag crosses a transcript splice junction. To avoid ambiguity, we used SAGE tags that mapped to a unique genomic position for further analysis (with 100% identity and no gaps). The mapping generated a set of 1,674 antisense SAGE tags overlapping with 1,351 sense transcripts (Tables **2** and supplementary Table **S1**). Of these tags, 71 were located within transcripts on both genomic strands and thus could potentially represent either a sense or antisense transcript. All data and mappings can be queried using various tools in StemBase [33] and can be displayed in their genomic context.

We classified the antisense tags by their position relative to the sense gene; in the 5'UTR, exons, introns, or the 3'UTR (Fig. **2**). Antisense tags were most commonly found in the 3'UTR of their corresponding sense gene and least commonly found in introns.

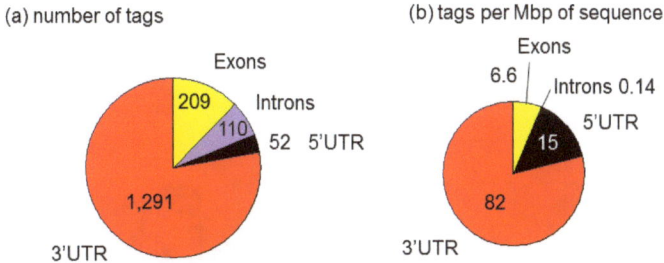

Figure 2: Location of antisense SAGE tags relative to sense gene. Statistics are for 1,674 tags overlapping 1,351 discrete Ensembl transcripts. Not shown, 6 tags overlap exon starts, 4 overlap exon ends and for 5 there was no information available of the gene exon/intron structure. Total genomic sequence content of exons (31 Mbp), introns (776 Mbp), 5'UTRs (3.5 Mbp) and 3'UTRs (16 Mbp) were estimated using 22,858 murine protein coding genes from Ensembl (version 42, January 2007). The majority of antisense tags were observed on the 3'UTRs of their corresponding sense gene. Normalized by sequence length, 3'UTRs are also the most rich in antisense tags, with introns having a 600-fold lower density (only 110 tags were found for a total length of 776 Mbp of introns in all murine genes).

Next, we compared the expression of the antisense tag to the expression state of the gene as measured in the microarray data. Overall, the sense gene was more often expressed if the antisense tag was present in the corresponding SAGE library (Fig. **3**). This was observed in all SAGE libraries we produced. When partitioned by position of the antisense tag within the gene, very low, or even negative, association was observed for the antisense tags in the introns, whereas positive correlation was strongest for antisense tags in the exons. Correlation between sense and antisense transcripts has been previously observed by Cawley *et al.* [34]. The authors propose that transcription factor binding sites found both 5' and 3' from genes may at least in part explain this co-expression.

We also observed this correlation at the protein level by comparing the antisense tag counts of the R1 mESC data with equivalent proteomics data [28] (see Methods). Of the 157 mouse proteins identified in R1 ES by Elliot *et al.*, 31 had associated antisense SAGE tags and for 30 of these 31 proteins the corresponding transcript producing the protein was identified as "Present" in the R1 Affymetrix data, indicating gene expression; thus the DNA microarray data confirms the expression of these proteins. Of 35 antisense tags overlapping the transcripts producing these proteins, 27 (77%) were observed in the R1 SAGE library, compared to 40% (677/1674) of all putative antisense tags. This suggests that if an antisense tag is detected the protein product of the sense gene is more likely to be produced (significance p-value < 1e-05 using the hypergeometric distribution).

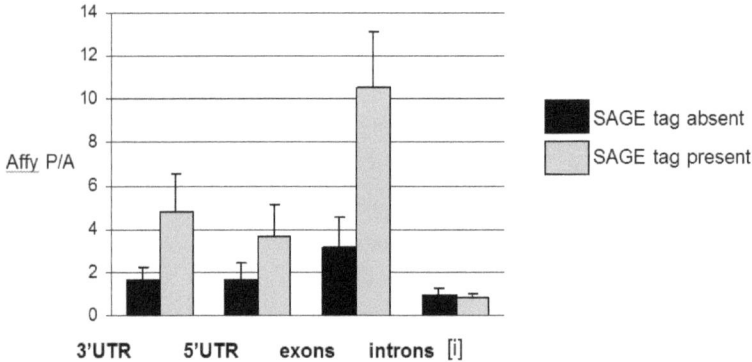

Figure 3: Ratios of expression of genes containing antisense tags. Average ratio of expressed to non-expressed genes (P/A, according to microarray data) containing an antisense SAGE tag in the region indicated on the x-axis. Black bars indicate expression ratio in the cell type where the SAGE tag is present, grey bars the ratio where the SAGE tag is absent (*i.e.* tag seen in a different SAGE library, but no evidence of antisense transcription in this cell type). P/A ratios are the mean of the ratios in four SAGE library/microarray comparisons (R1 ESC, R1 Embryoid bodies, Mammospheres, Differentiated mammospheres). Error bars indicate the standard deviation. The data used to compile this table are given as Supplementary Table **S1**.

We examined the expression pattern of our set of antisense tags in the Gene Atlas collection of SAGE libraries from 123 murine tissues [35]. The pattern of expression of sense transcripts detected by Affymetrix microarrays across 88 murine samples in StemBase is shown in Fig. **4a**. The distribution of the number of tissues in which a transcript is expressed is U-shaped, indicating that protein-coding genes are most commonly either tissue specific (expressed in only a few samples), or expressed in most tissues (likely to be 'housekeeping' genes). The expression pattern of antisense tags in the Gene Atlas libraries does not follow this pattern, instead showing a decreasing trend (Fig. **4b**, magenta points), indicating that antisense transcripts are more likely to be tissue specific and that there is little 'housekeeping' antisense transcription. By contrast, the pattern for all SAGE tags (sense and antisense) observed in the Gene Atlas dataset (Fig. **4b**, blue points), more closely resembles the pattern for Affymetrix probe sets in StemBase. This suggests that antisense transcripts are expressed in specific conditions, which is consistent with a function as modulators of a sense gene.

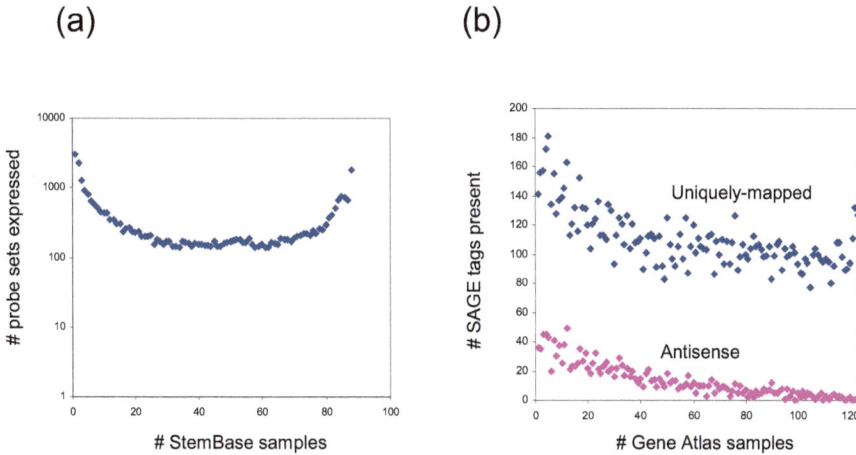

Figure 4: Ubiquity of expression measured by DNA microarrays and SAGE. (a) Expression pattern of Affymetrix probe sets across StemBase [31]. Points indicate the number of probe sets expressed ("present") in the number of samples indicated on the x-axis. Very few probe sets detect antisense transcripts, thus this distribution reflects sense transcript expression. These transcripts are most often either tissue specific or widely expressed (housekeeping). **(b)** Expression pattern of SAGE tags across Gene Atlas libraries [35]. Blue points represent all uniquely-mapped SAGE tags; magenta points represent our set of antisense tags (42 were absent from all 123 libraries examined). The distribution of all tags generally reflects (a), with a greater number of tags expressed in few or all tissues. By contrast the antisense tags have a continuously decreasing distribution as the number of tissues of expression increases.

One concern with the identification of antisense tags in a SAGE library is that the tags might be an artifact of library preparation, rather than representing a genuine antisense transcript. For 37% of the antisense tags we observed, the tag associated with the same *Nla*III site on the opposite (sense) strand was also seen in the library; this could indicate tag generation from an incorrectly primed or rearranged transcript, or from the non-anchored fragment generated by *Nla*III cleavage. If this is indeed the case, it could indicate strand errors similar to those seen in Expressed Sequence Tag generation. This observation does not guarantee that these tags are artifactual; they may still correspond to genuine antisense transcripts. It does, however, emphasize that the evidence for such antisense tags should be carefully weighed and additional evidence for antisense transcription sought wherever possible. By searching in dbEST we were able to identify for 735 of the 1,674 SAGE tags ESTs that were situated within 500nt of the genomic position of the tag and had a polyadenylation signal (PAS) within 30nt of the EST end in the same sense of the SAGE tag. This provides additional supporting evidence that the subset contains a large number of expressed antisense transcripts (Table **S2**).

The role of *cis*-acting antisense transcripts is unclear. One suggested role, in the control of the splicing of the sense gene, is under debate [36] although some examples have been verified. Alternative 3'UTR gene termination has been found to correlate to splice variants [37] and since most of the antisense transcripts we have identified overlap 3'UTRs it seemed reasonable to investigate whether there is an association between the antisense tags and alternative splicing. As described in the Methods, we identified genes with alternatively spliced forms and searched for association with our set of antisense tags. Our observations were not conclusive. The highest difference in ratios of spliced *versus* non-spliced genes (Table **3**; column 5) between expressed and non-expressed sense genes was found in mESC. Thus, we decided to examine mESC tags with differential expression during differentiation.

Table 3: Contingency tables for antisense tag expression *versus* sense gene splicing.

		Splice	No splice	Ratio
R1 ESC	exp	407	270	1.51
	no exp	556	441	1.26
R1 Embryoid bodies	exp	384	298	1.29
	no exp	579	413	1.40
Mammospheres	exp	541	396	1.37
	no exp	422	315	1.34
Differentiated mammospheres	exp	274	216	1.27
	no exp	689	495	1.39
Neural Stem Cells	exp	259	197	1.31
	no exp	704	514	1.37
Committed Neural Progenitors	exp	463	327	1.41
	no exp	500	384	1.30

Of the 1,674 antisense tags, 48 (antisense to 40 genes) had at least a five-fold difference in tag count between the stem cell and its differentiated counterpart in at least one of the three paired libraries (absent tags are given a tag count of one to avoid division by zero). From these we chose the 12 tags with differential expression in the R1 mESC differentiation for further study. Of these, 8 were more highly expressed in the mESC and in all cases the microarray data indicated that the sense gene is expressed both in the stem cell and in the differentiated state (Fig. **5**); these antisense tags may be hypothesized to have a function related to the maintenance of mESC pluripotency. We selected tags antisense to three genes for more detailed investigation: Zfp42/Rex1, Ywhag/14-3-3g and Prsp1 (Table **4**). For each of these genes we have compiled evidence of the gene's function and how it might be related to stem cell differentiation, expression patterns in stem cells and evidence supporting the presence of an antisense transcript (antisense ESTs containing a PAS).

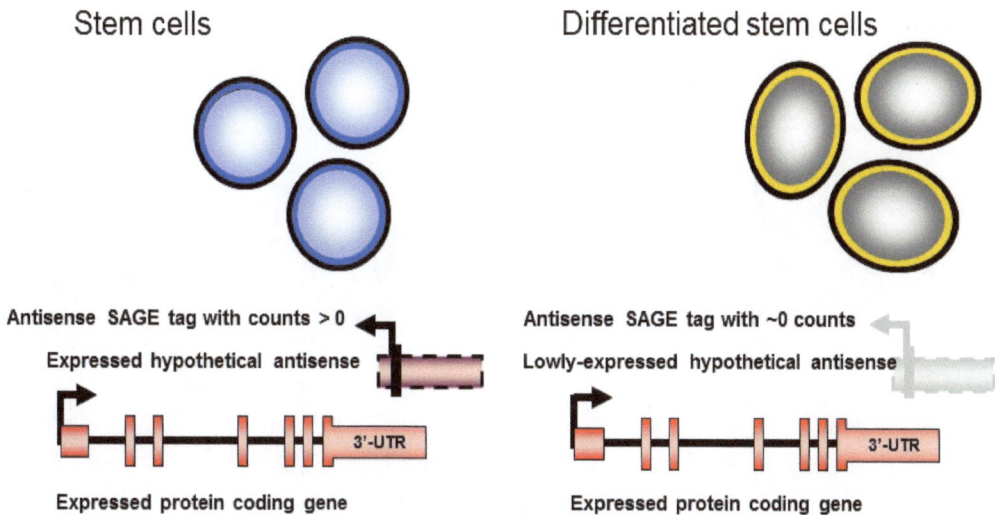

Figure 5: Model for selection of antisense tags differentially expressed in stem cell differentiation. SAGE tags that overlap known gene transcripts, but are on the opposite DNA strand, may represent longer antisense transcripts which could play a role in regulating expression of the sense gene. Antisense tags present in the stem cell library but absent in the corresponding differentiated library, may be hypothesized to have a function related to the maintenance of stem cell pluripotency. Expression of the sense gene in both samples suggests that the antisense transcript is acting as a modulator in the undifferentiated state. Our selection model involves identification of such transcripts, using other evidence to prioritize transcripts less likely to be artifacts of SAGE library generation.

Ywhag/14-3-3g encodes the gamma isoform of the tyrosine 3-monooxygenase/tryptophan 5-monooxygenase activation protein (14-3-3 protein). The family of 14-3-3 proteins regulates signal transduction by binding to serine/threonine phosphorylated proteins [38]. The gamma isoform binds to several proteins including actin [39] and signaling proteins such as BAD, RAF1, middle T antigen, BCR and PI-3 kinase [40]. Although this gene is highly expressed in some adult tissues, a developmental role has been demonstrated in *Xenopus* embryos, where morpholino antisense blocking of Ywhag produced eye defects and other abnormalities [41]. The distribution of expression in StemBase, according to MarkerServer [42], is mostly embryonic (Fig. **6a**); a number of other actin binding genes have a similar pattern of expression (MarkerServer pattern 761, multiple test corrected P-value = 0.02; genes Capzb, Ptk9, Tmsb10, Epb1.1l2, Cfl1, Macf1, Wdr1, Lasp1, Coro1c, Fscn1). We find numerous ESTs on the same strand as the SAGE tag, with multiple potential PAS. One in particular, 69nt downstream the tag, has 14 ESTs terminating within 30nt (EST BM250996 is one of them; Fig. (**7a**)).

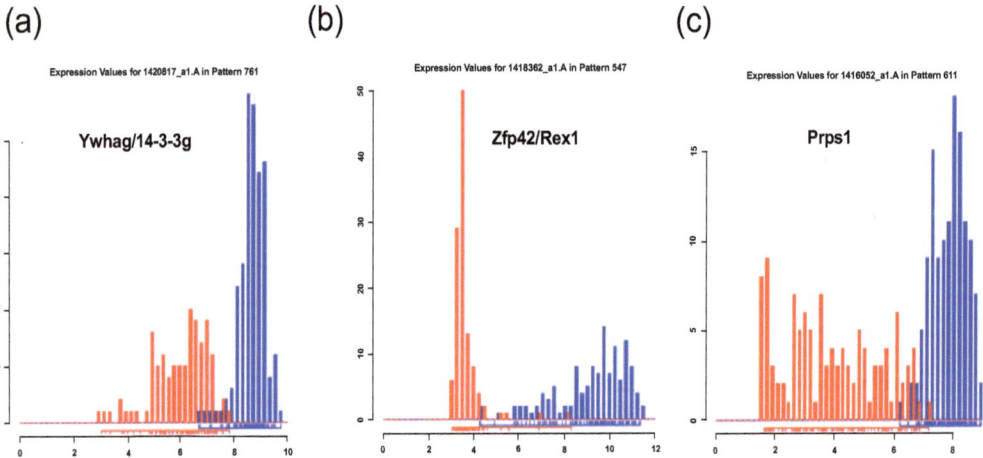

Figure 6: Expression of three genes in stem cells and derivatives. These figures are taken from MarkerServer [42], which contains pre-computed classifications of genes based on gap analysis of Affymetrix gene expression data in StemBase. The histograms show the number of samples in StemBase with a given hybridization value for the probe set identified in the legend. **(a)** Ywhag (probe set 1420812_at) is expressed at a high level in mESC lines (V6.5, R1, J1, C2 and D4, in blue). **(b)** Zfp42 (probe set 1418362_at) is expressed at a high level in mESC lines (V6.5, R1, J1, C2 and D4), samples of P19 EC cells, neurospheres and mammospheres (in blue). **(c)** Prps1 (probe set 1416052_at) is expressed at a high level in mESC lines (V6.5, R1, J1, C2 and D4) and in P19 EC cells (in blue). These graphs and the underlying data can be obtained from MarkerServer using the probe set identifiers given above (*e.g.* 1420812_at for Ywhag).

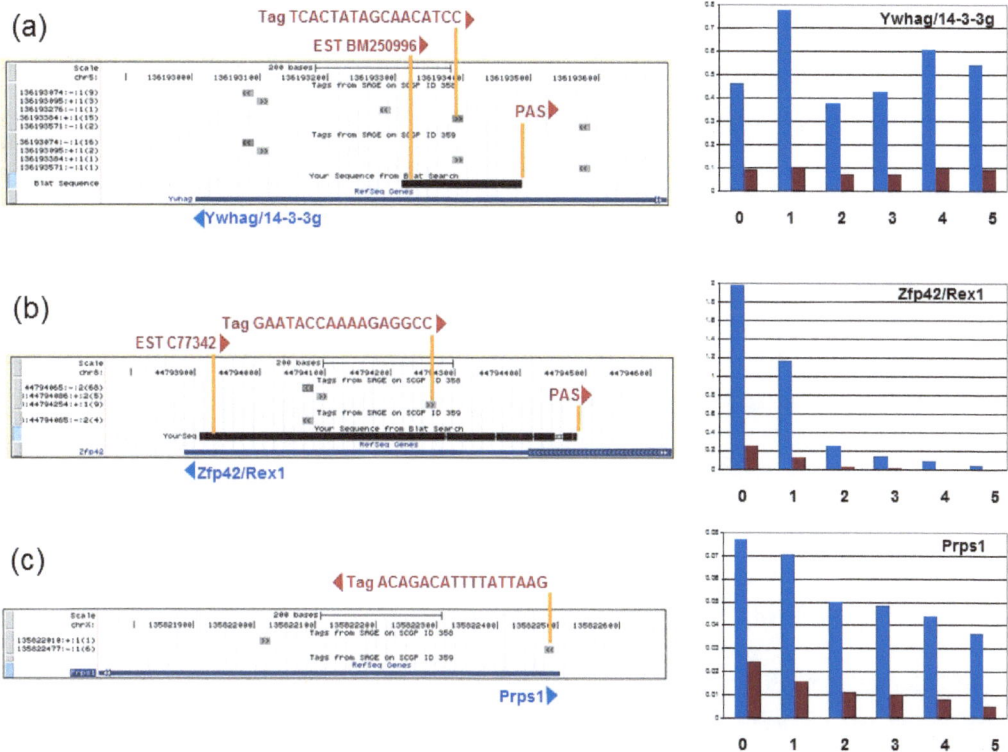

Figure 7: Sense and antisense transcription for three genes. Left panels: SAGE data was overlaid to the UCSC Genome view using the StemBase genomic viewer [33]. Upper tracks (gray boxes) indicate the position of SAGE tags with labels indicating genomic position, strand, number of positions mapped in the genome and tag count (in brackets). Blue boxes are RefSeq genes, black boxes (if present) represent ESTs supporting antisense transcript. Right panels: RT-qPCR analysis of sense and corresponding putative antisense transcripts in 5-day differentiation of R1 mESC cells. PCR Primers are given in Table **5**. Data is for **(a)** Zfp42, **(b)** Ywhag and **(c)** Prps1. Blue and red bars represent sense and antisense transcripts respectively.

Zfp42/Rex1, zinc finger protein 42, is a transcription factor that has been identified as an embryonic stem cell marker. However, its function remains unknown and whereas some studies associate it to differentiation [43] others find it not to be crucial for pluripotency [44]. The distribution of Zfp42 expression in StemBase according to MarkerServer agrees with its designation as a stem cell marker (Fig. **6b**). This is one of 9 genes with a heterogeneous pattern of expression at single cell level, demonstrated using *in situ* hybridization in ES cells [45]; this suggests a complex role for Zfp42 in stem cell differentiation. The SAGE tag overlaps with an

EST on the same strand, containing a PAS 202nt downstream from the SAGE tag (EST C77342; Fig. (**7b**)). We observed no sense tag derived from the same NlaIII site, providing support that this might represent a real antisense transcript.

Prps1 encodes the phosphoribosyl pyrophosphate synthetase 1, a protein that catalyzes the synthesis of phosphoribosyl pyrophosphate (PRPP) from ATP and ribose 5-phosphate [46]. The expression patterns of this gene, in humans expressed in all tissues, suggest that it has a housekeeping role. However, it is highly expressed in mESC according to MarkerServer (Fig. **6c**). In humans, the abnormally increased expression of this gene occurs in an inherited X-linked disorder associated to uric acid overproduction and neurodevelopmental abnormalities [47] and loss of function mutations have been associated to neural diseases [48, 49]. We could find no EST information supporting this antisense tag (Fig. **7c**). However, we observed no sense tag derived from the same *Nla*III site, supporting the possibility that this represents a real antisense transcript.

Table 4: Three differentially expressed antisense tags.

SAGE tag	Chr	Strand	Gene Symbol	Tag counts					
				R1 ESC	R1 Embryoid bodies	Mammo -spheres	Differentiated Mammo- spheres	Neural Stem Cells	Committed Neural Progenitors
TCACTAT AGCAACA TCC	5	+	Ywhag	15	1	8	0	3	0
GAATACC AAAAGAG GCC	8	+	Zfp42	9	0	0	0	0	0
ACAGACA TTTTATT AAG	X	-	Prps1	6	0	0	0	0	0

Based on the position of the antisense SAGE tags and of the prediction of the gene (Table **4**), we designed specific primers for a transcript expected to produce the corresponding antisense tag (Table **5**). The tool PDA (Primer Design Assistant; [50]) was used for primer design. RT-qPCR was used to assess the expression of sense and antisense transcripts during five days of differentiation of R1 mESC (Fig. **7**; right panels). For Zfp42 and Prps1, the sense transcript decreases during differentiation, preceded by a decrease in the antisense transcript. The pattern in the case of Ywhag is more complex. However, at all time points and cases analysed, the variation in the sense gene expression is greater than that of the antisense transcript.

Table 5: RT-qPCR Primers.

Ywhag sense	AGCAACATCCAAAACAGATCAA
Ywhag antisense	CGGTAGCTTGACATTTATTACCTT
Prps1 sense	ATGTGCTGGCCAGTGAAAC
Prps1 antisense	ATCTTCCTTGGATCCGTTTG
Zfp42 sense	CGTATGCAAAAGTCCCCATC
Zfp42 antisense	TGTCCTCAGGCTGGGTAGTC

These three genes all have evidence of alternative splicing, according to predictions from the Swiss Institute of Bioinformatics (SIB) (data not shown) and from splicing models in AceView and Alt Events in the UCSC Genome Browser. However, we could not find evidence of these alternate splice forms in the literature. Antisense transcription has been implicated in control of splice variation of the sense genes (see *e.g.* [51]). Given the weak association between antisense SAGE tags and genes with alternate splice forms in the mESC library (Table **3**) we wondered if we would be able to identify novel splicing events in the RNAs encoded by these genes.

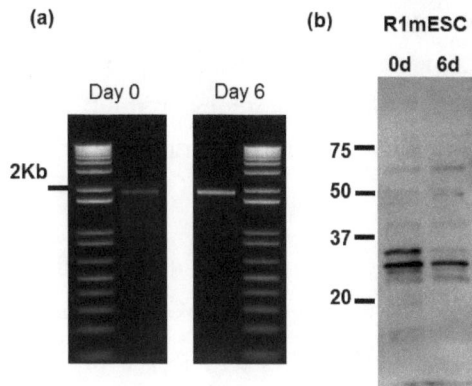

Figure 8: Study of Prps1 transcript and protein products in R1 mESC differentiation. (a) RT-qPCR of Prps1 displaying a single 1.9kb transcript on day 0 and day 6. **(b)** Western blot of PRPS1. Left lane, undifferentiated R1 mESC. Right lane, differentiated R1 mESC (6 days), 5μg of total protein were loaded in each lane. The higher molecular weight band corresponds to the native form of the protein (35kDa). A lower molecular weight form is detected that remains in the differentiated state.

We chose Prps1 to test this approach, looking for splice variants at both the transcript and protein level pre- and post-differentiation. We searched for variation in the transcript by RT PCR; this detected only the expected 1.9 kb product (Fig. **8a**). A Western blot using an antibody against the product of Prps1 displayed the primary band of the protein and a secondary band of very similar weight that was not consistent with the change in molecular weight expected from the SIB predictions. We interpret this second band as a post-translationally modified version of the gene (5kDa lower molecular weight, Fig. **8b**). If the transcript antisense to Prps1 has any effect on the expression of the gene in mESC differentiation this evidence suggests that this will likely not involve splice variation.

CONCLUSION

We identified evidence of antisense transcription in a number of SAGE libraries generated from mouse stem cells and their differentiated derivatives; this finding is supported by a number of other studies that have identified antisense SAGE tags in mammals and other organisms. In six SAGE libraries we identified 1,674 antisense SAGE tags, which overlap with 1,351 sense transcripts; these sense/antisense transcript pairs represent potential regulatory relationships that warrant further investigation. To prioritize candidates for validation, we identified 48 antisense tags that vary in abundance between undifferentiated and differentiated cells. These tags identify 40 genes that are candidates for antisense regulation during the stem cell differentiation process. Microarray evidence indicates that the majority of these genes remain expressed through the differentiation process, suggesting a modulatory rather than on/off mode for antisense regulation.

We verified the expression of the gene and the associated antisense transcript for three tags with decreasing expression in mESC differentiation. Our results for these three tags suggest that they represent antisense transcripts that may regulate genes with functions related to stem cells. Our data constitute a repository of paired SAGE and microarray data that can be used to identify antisense transcripts acting as regulators of stem cell differentiation and supports the idea that antisense transcription plays a yet undiscovered role in many aspects of biological control.

REFERENCES

[1]   Velculescu VE, Zhang L, Vogelstein B, Kinzler KW: Serial analysis of gene expression. *Science* 1995, 270(5235):484-487.

[2]   Barrett T, Troup DB, Wilhite SE, Ledoux P, Rudnev D, Evangelista C, Kim IF, Soboleva A, Tomashevsky M, Marshall KA *et al*: NCBI GEO: archive for high-throughput functional genomic data. *Nucleic Acids Res* 2009, 37(Database issue):D885-890.

[3]   Wang Z, Gerstein M, Snyder M: RNA-Seq: a revolutionary tool for transcriptomics. *Nat Rev Genet* 2009, 10(1):57-63.

[4]   Poole RL, Barker GL, Werner K, Biggi GF, Coghill J, Gibbings JG, Berry S, Dunwell JM, Edwards KJ: Analysis of wheat SAGE tags reveals evidence for widespread antisense transcription. *BMC Genomics* 2008, 9:475.

[5]   Calsa T, Jr., Figueira A: Serial analysis of gene expression in sugarcane (Saccharum spp.) leaves revealed alternative C4 metabolism and putative antisense transcripts. *Plant Mol Biol* 2007, 63(6):745-762.

[6]   Gibbings JG, Cook BP, Dufault MR, Madden SL, Khuri S, Turnbull CJ, Dunwell JM: Global transcript analysis of rice leaf and seed using SAGE technology. *Plant Biotechnol J* 2003, 1(4):271-285.

[7]   Robinson SJ, Cram DJ, Lewis CT, Parkin IA: Maximizing the efficacy of SAGE analysis identifies novel transcripts in Arabidopsis. *Plant Physiol* 2004, 136(2):3223-3233.

[8]   Gowda M, Venu RC, Raghupathy MB, Nobuta K, Li H, Wing R, Stahlberg E, Couglan S, Haudenschild CD, Dean R *et al*: Deep and comparative analysis of the mycelium and appressorium transcriptomes of Magnaporthe grisea using MPSS, RL-SAGE and oligoarray methods. *BMC Genomics* 2006, 7:310.

[9]   Lee S, Bao J, Zhou G, Shapiro J, Xu J, Shi RZ, Lu X, Clark T, Johnson D, Kim YC *et al*: Detecting novel low-abundant transcripts in Drosophila. *RNA* 2005, 11(6):939-946.

[10]   Gunasekera AM, Patankar S, Schug J, Eisen G, Kissinger J, Roos D, Wirth DF: Widespread distribution of antisense transcripts in the Plasmodium falciparum genome. *Mol Biochem Parasitol* 2004, 136(1):35-42.

[11]   Jones SJ, Riddle DL, Pouzyrev AT, Velculescu VE, Hillier L, Eddy SR, Stricklin SL, Baillie DL, Waterston R, Marra MA: Changes in gene expression associated with developmental arrest and longevity in Caenorhabditis elegans. *Genome Res* 2001, 11(8):1346-1352.

[12]   Chen J, Sun M, Kent WJ, Huang X, Xie H, Wang W, Zhou G, Shi RZ, Rowley JD: Over 20% of human transcripts might form sense-antisense pairs. *Nucleic Acids Res* 2004, 32(16):4812-4820.

[13]   Keime C, Semon M, Mouchiroud D, Duret L, Gandrillon O: Unexpected observations after mapping LongSAGE tags to the human genome. *BMC Bioinformatics* 2007, 8:154.

[14]   Ge X, Wu Q, Jung YC, Chen J, Wang SM: A large quantity of novel human antisense transcripts detected by LongSAGE. *Bioinformatics* 2006, 22(20):2475-2479.

[15]   Richards M, Tan SP, Chan WK, Bongso A: Reverse serial analysis of gene expression (SAGE) characterization of orphan SAGE tags from human embryonic stem cells identifies the presence of novel transcripts and antisense transcription of key pluripotency genes. *Stem Cells* 2006, 24(5):1162-1173.

[16]   Quere R, Manchon L, Lejeune M, Clement O, Pierrat F, Bonafoux B, Commes T, Piquemal D, Marti J: Mining SAGE data allows large-scale, sensitive screening of antisense transcript expression. *Nucleic Acids Res* 2004, 32(20):e163.

[17]   Shendure J, Church GM: Computational discovery of sense-antisense transcription in the human and mouse genomes. *Genome Biol* 2002, 3(9):RESEARCH0044.

[18]   Wahl MB, Heinzmann U, Imai K: LongSAGE analysis significantly improves genome annotation: identifications of novel genes and alternative transcripts in the mouse. *Bioinformatics* 2005, 21(8):1393-1400.

[19]   Lavorgna G, Dahary D, Lehner B, Sorek R, Sanderson CM, Casari G: In search of antisense. *Trends Biochem Sci* 2004, 29(2):88-94.

[20]   Katayama S, Tomaru Y, Kasukawa T, Waki K, Nakanishi M, Nakamura M, Nishida H, Yap CC, Suzuki M, Kawai J *et al*: Antisense transcription in the mammalian transcriptome. *Science* 2005, 309(5740):1564-1566.

[21]   Sleutels F, Zwart R, Barlow DP: The non-coding Air RNA is required for silencing autosomal imprinted genes. *Nature* 2002, 415(6873):810-813.

[22]   Zhang Y, Liu XS, Liu QR, Wei L: Genome-wide *in silico* identification and analysis of *cis* natural antisense transcripts (*cis*-NATs) in ten species. *Nucleic Acids Res* 2006, 34(12):3465-3475.

[23]   Chen J, Sun M, Hurst LD, Carmichael GG, Rowley JD: Genome-wide analysis of coordinate expression and evolution of human *cis*-encoded sense-antisense transcripts. *Trends Genet* 2005, 21(6):326-329.

[24]   Kapranov P, Willingham AT, Gingeras TR: Genome-wide transcription and the implications for genomic organization. *Nat Rev Genet* 2007, 8(6):413-423.

[25]   Mercer TR, Dinger ME, Sunkin SM, Mehler MF, Mattick JS: Specific expression of long noncoding RNAs in the mouse brain. *Proc Natl Acad Sci USA* 2008, 105(2):716-721.

[26]   Ebralidze AK, Guibal FC, Steidl U, Zhang P, Lee S, Bartholdy B, Jorda MA, Petkova V, Rosenbauer F, Huang G *et al*: PU.1 expression is modulated by the balance of functional sense and antisense RNAs regulated by a shared *cis*-regulatory element. *Genes Dev* 2008, 22(15):2085-2092.

[27]   Hailesellasse Sene K, Porter CJ, Palidwor G, Perez-Iratxeta C, Muro EM, Campbell PA, Rudnicki MA, Andrade-Navarro MA: Gene function in early mouse embryonic stem cell differentiation. *BMC Genomics* 2007, 8:85.

[28]   Elliott ST, Crider DG, Garnham CP, Boheler KR, Van Eyk JE: Two-dimensional gel electrophoresis database of murine R1 embryonic stem cells. *Proteomics* 2004, 4(12):3813-3832.

[29]   Thierry-Mieg D, Thierry-Mieg J: AceView: a comprehensive cDNA-supported gene and transcripts annotation. *Genome Biol* 2006, 7 Suppl 1:S12 11-14.

[30]   Kent WJ, Sugnet CW, Furey TS, Roskin KM, Pringle TH, Zahler AM, Haussler D: The human genome browser at UCSC. *Genome Res* 2002, 12(6):996-1006.

[31]   Perez-Iratxeta C, Palidwor G, Porter CJ, Sanche NA, Huska MR, Suomela BP, Muro EM, Krzyzanowski PM, Hughes E, Campbell PA *et al*: Study of stem cell function using microarray experiments. *FEBS Lett* 2005, 579(8):1795-1801.

[32]   Beissbarth T, Hyde L, Smyth GK, Job C, Boon WM, Tan SS, Scott HS, Speed TP: Statistical modeling of sequencing errors in SAGE libraries. *Bioinformatics* 2004, 20 Suppl 1:I31-I39.

[33]   Sandie R, Palidwor GA, Huska MR, Porter CJ, Krzyzanowski PM, Muro EM, Perez-Iratxeta C, Andrade-Navarro MA: Recent developments in StemBase: a tool to study gene expression in human and murine stem cells. *Submitted* 2009.

[34] Cawley S, Bekiranov S, Ng HH, Kapranov P, Sekinger EA, Kampa D, Piccolboni A, Sementchenko V, Cheng J, Williams AJ *et al*: Unbiased mapping of transcription factor binding sites along human chromosomes 21 and 22 points to widespread regulation of noncoding RNAs. *Cell* 2004, 116(4):499-509.

[35] Siddiqui AS, Khattra J, Delaney AD, Zhao Y, Astell C, Asano J, Babakaiff R, Barber S, Beland J, Bohacec S *et al*: A mouse atlas of gene expression: large-scale digital gene-expression profiles from precisely defined developing C57BL/6J mouse tissues and cells. *Proc Natl Acad Sci USA* 2005, 102(51):18485-18490.

[36] Werner A, Carlile M, Swan D: What do natural antisense transcripts regulate? *RNA Biol* 2009, 6(1):43-48.

[37] Wang ET, Sandberg R, Luo S, Khrebtukova I, Zhang L, Mayr C, Kingsmore SF, Schroth GP, Burge CB: Alternative isoform regulation in human tissue transcriptomes. *Nature* 2008, 456(7221):470-476.

[38] Aitken A: 14-3-3 proteins: a historic overview. *Semin Cancer Biol* 2006, 16(3):162-172.

[39] Chen XQ, Yu AC: The association of 14-3-3gamma and actin plays a role in cell division and apoptosis in astrocytes. *Biochem Biophys Res Commun* 2002, 296(3):657-663.

[40] Zha J, Harada H, Yang E, Jockel J, Korsmeyer SJ: Serine phosphorylation of death agonist BAD in response to survival factor results in binding to 14-3-3 not BCL-X(L). *Cell* 1996, 87(4):619-628.

[41] Lau JM, Muslin AJ: Analysis of 14-3-3 family member function in Xenopus embryos by microinjection of antisense morpholino oligos. *Methods Mol Biol* 2009, 518:31-41.

[42] Krzyzanowski PM, Andrade-Navarro MA: Identification of novel stem cell markers using gap analysis of gene expression data. *Genome Biol* 2007, 8(9):R193.

[43] Thompson JR, Gudas LJ: Retinoic acid induces parietal endoderm but not primitive endoderm and visceral endoderm differentiation in F9 teratocarcinoma stem cells with a targeted deletion of the Rex-1 (Zfp-42) gene. *Mol Cell Endocrinol* 2002, 195(1-2):119-133.

[44] Masui S, Ohtsuka S, Yagi R, Takahashi K, Ko MS, Niwa H: Rex1/Zfp42 is dispensable for pluripotency in mouse ES cells. *BMC Dev Biol* 2008, 8:45.

[45] Carter MG, Stagg CA, Falco G, Yoshikawa T, Bassey UC, Aiba K, Sharova LV, Shaik N, Ko MS: An *in situ* hybridization-based screen for heterogeneously expressed genes in mouse ES cells. *Gene Expr Patterns* 2008, 8(3):181-198.

[46] Kornberg A, Lieberman I, Simms ES: Enzymatic synthesis and properties of 5-phosphoribosylpyrophosphate. *J Biol Chem* 1955, 215(1):389-402.

[47] Roessler BJ, Nosal JM, Smith PR, Heidler SA, Palella TD, Switzer RL, Becker MA: Human X-linked phosphoribosylpyrophosphate synthetase superactivity is associated with distinct point mutations in the PRPS1 gene. *J Biol Chem* 1993, 268(35):26476-26481.

[48] Kim HJ, Sohn KM, Shy ME, Krajewski KM, Hwang M, Park JH, Jang SY, Won HH, Choi BO, Hong SH *et al*: Mutations in PRPS1, which encodes the phosphoribosyl pyrophosphate synthetase enzyme critical for nucleotide biosynthesis, cause hereditary peripheral neuropathy with hearing loss and optic neuropathy (cmtx5). *Am J Hum Genet* 2007, 81(3):552-558.

[49] de Brouwer AP, Williams KL, Duley JA, van Kuilenburg AB, Nabuurs SB, Egmont-Petersen M, Lugtenberg D, Zoetekouw L, Banning MJ, Roeffen M *et al*: Arts syndrome is caused by loss-of-function mutations in PRPS1. *Am J Hum Genet* 2007, 81(3):507-518.

[50] Chen SH, Lin CY, Cho CS, Lo CZ, Hsiung CA: Primer Design Assistant (PDA): A web-based primer design tool. *Nucleic Acids Res* 2003, 31(13):3751-3754.

[51] Scheele C, Petrovic N, Faghihi MA, Lassmann T, Fredriksson K, Rooyackers O, Wahlestedt C, Good L, Timmons JA: The human PINK1 locus is regulated *in vivo* by a non-coding natural antisense RNA during modulation of mitochondrial function. *BMC Genomics* 2007, 8:74.

[52] Liu G, Loraine AE, Shigeta R, Cline M, Cheng J, Valmeekam V, Sun S, Kulp D, Siani-Rose MA: NetAffx: Affymetrix probesets and annotations. *Nucleic Acids Res* 2003, 31(1):82-86.

[53] Porter CJ, Palidwor GA, Sandie R, Krzyzanowski PM, Muro EM, Perez-Iratxeta C, Andrade-Navarro MA: StemBase: a resource for the analysis of stem cell gene expression data. *Methods Mol Biol* 2007, 407:137-148.

SUPPLEMENTARY MATERIAL

Two supplementary tables are available as supplementary material from http://www.ogic.ca/projects/affysage/sandie_affysage_2012.zip

Table S1: Pairs of antisense SAGE tags and Affymetrix probe sets for related sense transcript.

A total of 2,910 pairs of SAGE tags and Affymetrix probe sets describing expression of an overlapping sense transcript were generated. These correspond to 1,351 Ensembl sense transcripts (2,283 probe sets) and 1,674 SAGE antisense tags. 87 tags were differentially expressed between stem cell and derivative cell for at least one the three library pairs, overlapping 40 sense genes. Columns are (1) SAGE: SAGE tag, (2) Chr: chromosome to which the tag maps, (3, 4) From, To: start and end coordinates of the tag, (5) Strand: strand to which the tag maps, (6) Loc: position of the tag relative to the sense transcript: [5] 5'UTR, [Ex] exon, [I] intron, [3] 3'UTR, [S] overlap with start of exon, [E] overlap with end of exon, [n/a] no exon/intron information available, (7) TransID: Ensembl transcript identifier, (8) Gene Symbol, (9) Probe: probe set identifier for the Ensembl transcript. Following columns give the expression in SAGE libraries (number of counts) or Affymetrix experiments (transcript Present or otherwise Absent). StemBase sample identifier indicated within brackets, see details about samples in Table 1: (10) ES-SAGE(S358), (11) ES-Affy(S206), (12) EB-SAGE(S222), (13) EB-Affy(S216), (14) M-SAGE(S355), (15) M-Affy(S255), (16) DM-SAGE(S356), (17) DM-Affy(S256), (18) NS-SAGE(S331), (19) NS-Affy(S206), (20) CNP-SAGE(S332). Last two columns indicate expression in large databases: (21) gene atlas, number of Gene Atlas libraries where the SAGE tag was present (maximum possible 123), (22) StemBase, number of StemBase samples where the probe set was expressed (maximum possible 88).

Table S2: Antisense SAGE tags to mouse sense transcripts and EST evidence.

Supporting EST evidence for 1,608 antisense SAGE tags that do not overlap a coding transcript on the same strand as the SAGE tag. Columns (1) position of the tag relative to the sense transcript, (2) SAGE tag, (3) chromosome where the tag was mapped, (4) start coordinate, (5) end coordinate, (6) strand to which the tag maps, (7) distance and number of ESTs supporting a nearby antisense PAS (multiple PAS are possible). A total of 788 tags have EST support.

CHAPTER 12

Computational Analysis of ChIP-seq Data and Its Application to Embryonic Stem Cells

Xu Han and Lin Feng[*]

Nanyang Technological University, Nanyang Avenue, Singapore

Abstract: With the advent of ultra-high-throughput sequencing technologies, ChIP-seq is becoming the main stream for the genome-wide studies of transcription factor binding sites (TFBSs) and histone modification sites. Computational analysis of ChIP-seq data is important for ChIP-seq applications. In this chapter, we first give an overview of the state-of-the-art ChIP-seq analysis tools developed for predicting ChIP-enriched genomic sites. Next, we describe the methods employed in a comprehensive analysis on Chen *et al.*'s ChIP-seq dataset in mouse embryonic stem cells (mESC) [1]. These methods include the prediction of transcription factor binding peaks, as well as subsequent analysis procedures such as *de novo* motif-finding and the discovery of transcription factor co-localization. By this, we demonstrate how the computational approaches assist to achieve novel biological discoveries from large-scale ChIP-seq dataset.

Keywords: Computational Analysis, Sequencing, Transcriptional regulation, Chromatin Immuno-precipitation, Histone modification, ChIP-seq, ChIP-enriched loci, Hidden Markov Model, Control library, Poisson distribution, mESC.

INTRODUCTION

In early 2007, Illumina Inc. launched the Solexa 1G genome analyzer, the milestone of the next generation ultra-high-throughput sequencing technologies. In subsequent years, the sequencing capacity has been tremendously improved, which led to genome-wide studies in various areas such as genome re-sequencing and SNP association [2], mRNA expression [3], DNA methylation [4], and 3-dimensional DNA structure [5]. One of the major (also the earliest) applications of the ultra-high-throughput sequencing technologies is ChIP-seq, in which DNA sequencing is coupled with Chromatin Immuno-precipitation (ChIP) experiment to identify histone modification sites (HMSs) or transcription factor binding sites

*Address correspondence to Lin Feng:** Nanyang Technological University, School of Computer Engineering, N4-2A-05, Nanyang Avenue, Singapore 639798; Tel: (65) 67906184; Fax: (65) 67926559; E-mail: asflin@ntu.edu.sg

(TFBSs). Barski *et al.* first employed ChIP-seq approach to profile 20 histone lysine and arginine methylations in human CD4$^+$ T cell [6]. Johnson *et al.* further adapted the ChIP-seq protocol to determine the TFBSs of NRSF in Jurkat T cell [7]. Research work by these two groups demonstrated the impressive power of ChIP-seq for the study of histone modification and transcription factor binding, in term of sensitivity, specificity and resolution.

Shortly after the advent of ChIP-seq, this technology was introduced to the area of embryonic stem cell research. Mikkelsen *et al.* used ChIP-seq to generate the profiles of several histone modifications, including H3K4me3, H3K36me3, H3K27me3, H3K9me3, in three different mouse cell types: embryonic stem cells (ESC), neural progenitor cells (NPCs) and embryonic fibroblasts (MEFs) [8]. Chen *et al.* produced ChIP-seq data for 15 sequence specific transcription factors or regulators in mESC [1]. These large-scale datasets provide important sources for better understanding of embryonic stem cell by the community.

Computational analysis of ChIP-seq data involves the identification of ChIP-enriched genomic loci and the post-analysis methods such as *de novo* motif-finding and multiple library comparison. In this chapter, we first give an overview of the state-of-the-art ChIP-seq analysis tools developed for predicting ChIP-enriched loci. Next, we describe the methods employed in a comprehensive analysis of Chen *et al.*'s ChIP-seq dataset on mouse embryonic stem cells (mESC) [1]. These methods include the prediction of transcription factor binding peaks, as well as subsequent analysis procedures such as *de novo* motif-finding and the determination of transcription factor co-localization. By this, we demonstrate how computational approaches assist to achieve novel biological discoveries from large-scale ChIP-seq dataset.

TOOLS FOR PREDICTING CHIP-ENRICHED SITES

In the ChIP-seq protocol, DNA fragments from the loci of interest (*i.e.,* TFBSs or HMSs) are enriched by an antibody following the ChIP procedure. The basic idea of ChIP-seq is to read the sequence of one end of a ChIP-enriched DNA fragment, and to map the short read to the genome assembly in order to find the genomic location of that fragment. Millions of reads sequenced from a ChIP library are mapped and form a genome-wide profile in which counts of read are overrepresented at the loci of interest. The patterns of ChIP-seq profile may vary among different applications (Fig. **1a**). For most ChIP-seq datasets on transcription factors, the profile appears sharp peaks centered by the actual binding sites; while for some datasets on histone modifications, much boarder patterns were observed in profiling, suggesting the enzyme that modifies the histone protein may affect a number of consecutive nucleosomes. Therefore, various application-dependent tools have been developed for identifying ChIP-enriched sites.

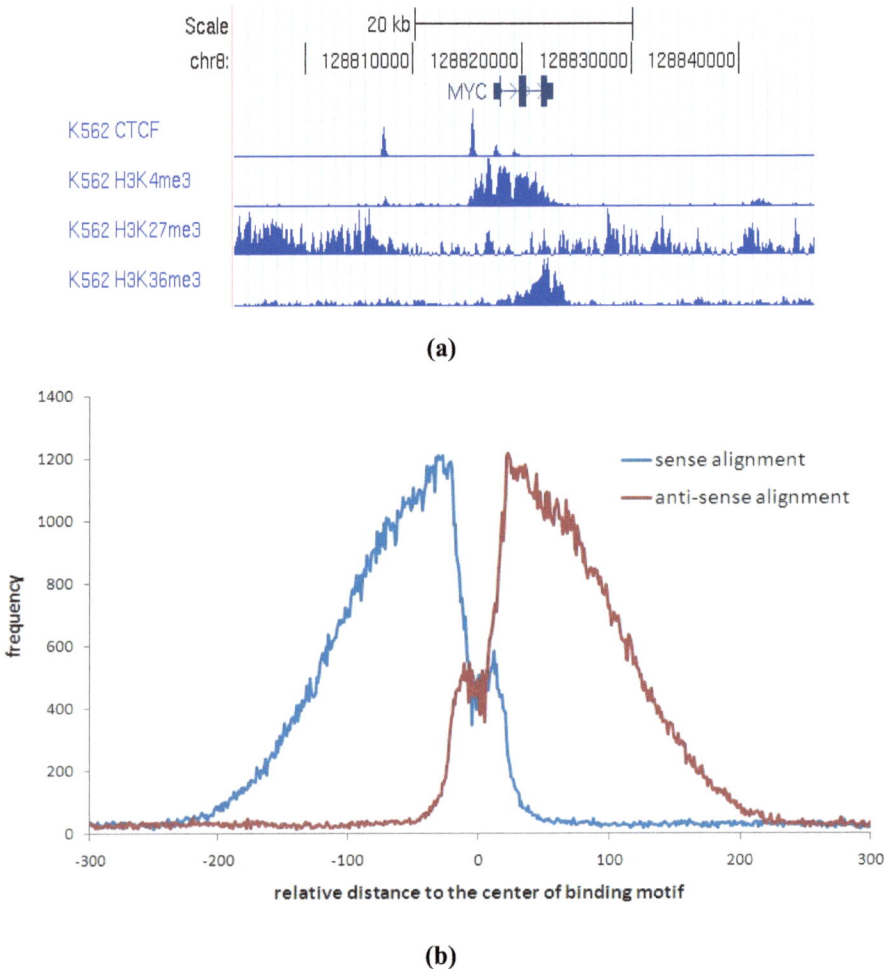

(a)

(b)

Figure 1: (a) Different patterns of ChIP-seq profile for CTCF, H3K4me3, H3K27me3, and H3K36me3 in proximal region of MYC gene in human K562 cell-line. **(b)** Aligned reads are directional and are oriented towards the actual binding sites. Data retrieved from the CTCF library in Chen *et al.*'s dataset.

The first software tool for ChIP-seq analysis on TFBS identification was introduced by Johnson *et al.* [7]. They employed a straight-forward algorithm to locate the summit of the peaks in the profile. The variation of read counts due to the random sampling of fragments could be further smoothed using Gaussian density kernel, which was implemented in the tool of FSeq [9]. Since a ChIP-seq read is sequenced from one end of a ChIP-enriched DNA fragment, the alignment to the genome assembly is directional and can be in either sense or anti-sense direction. Indeed, the aligned read is oriented towards the actual binding site (Fig. **1b**). In the light of this, several subsequent tools utilized the orientation information of reads to pursue a higher resolution of binding site prediction. FindPeaks [10] extended the reads towards there orientations to form "virtual fragments". Overlapping "virtual fragments" were accumulated into ChIP-seq profile, which better represents the binding peaks with improved resolution. QuEST [11] generated density profile for sense and anti-sense reads individually, and determined the binding sites as the center between sense and anti-sense peaks. SISSRs [12] predicted the binding sites to be transition position from sense reads to anti-sense reads. As reported in their corresponding publications, these approaches improved the resolution of peak prediction (measured as the median distance between predicted peaks to binding motif) from ~50 bps to ~20 bps.

Unlike those for transcription factor study, the ChIP-seq applications to histone modifications are mostly designed to reveal the ChIP-enriched genomic regions that may span from hundreds of bps to ~100k bps. Therefore, the predicted histone modification sites need to be defined in term of regions rather than individual peaks. Mikkelsen *et al.* proposed a sliding-window based approach to predict histone modification sites [8]. In each step of sliding, the significance of read counts in the window was tested against random Poisson model. They also developed an alternative approach based on hidden Markov model (HMM), which is more suitable to handle larger continuous regions. Zang *et al.* defined "islands" as small ChIP-enriched bins and clustered the proximal islands into larger regions [13]. This idea was implemented in SICER, which seems to be less sensitive to the bin size in comparison to Mikkelsen's windowing approach.

A common question in ChIP-seq analysis is how to control the specificity of prediction, which is usually measured by false discovery rate (FDR). In early

ChIP-seq application without a negative control, a random model of uniform distribution of DNA fragments was assumed for significance analysis. In this case, the expected number of random peaks can be estimated either from known distributions (*e.g.,* Poisson distribution [10], Negative binomial distribution [14]) or from randomly simulated dataset [1]. The FDR was then computed to be the ratio of the expected number of random peaks to the number of observed peak above certain threshold. However, recent studies showed that the uniform random model may not work well in practice due to the impact of sequencing and mapping biases, chromatin structure and genome copy number variations [15] [16]. Therefore, it has been suggested that a negative control library, which could be generated using non-specific antibody or input DNA, is required in a ChIP-seq experiment to reflect these intrinsic biases.

Table 1: A summary of ChIP-seq analysis tools for prediction ChIP-enriched sites.

Name or description of tool	Site representation	Significance measurement	With control library	Refs.
Johnson *et al.,* 's peak detection	peaks	read counts	no	[7]
Fseq	peaks	kernel density estimates	no	[9]
FindPeaks	peaks	Poission p-value	no	[10]
QuEST	peaks	kernel density estimates	yes/no	[11]
SISSR	peaks	read counts	no	[12]
Mikkelsen *et al.,* 's sliding-window approach	regions	read counts	no	[8]
Mikkelsen *et al.,* 's HMM approach	regions	probability	no	[8]
SICER	regions	island score	yes/no	[13]
MACs	peaks	Poisson p-value	yes	[15]
CisGenome (one-sample)	regions	negative binomial p-value	no	[14]
CisGenome (two-samples)	regions	binomial p-value	yes	[14]
PeakSeq	regions	binomial p-value	yes	[17]
GLITR	regions	number of nearest neighbor in 2D space	yes	[18]
ChIPDiff	regions	probability	yes*	[19]

*ChIPDiff compares two libraries generated from different cell-lines with the same antibody.

The most straight-forward approach to utilize the control library is to set an empirical threshold of fold-change against control, in order to filter out the falsely predicted peaks introduced due to intrinsic biases. However, fold-change measurement could be too sensitive to the random sampling variation when the sequencing is not deep enough. Therefore, several other tools performed statistical tests to determine the significant peaks. In MACs, a random model of Poisson distribution was assumed, where the parameter of Poisson distribution is a variable and was estimated from the read counts in local region from the control library [15]. CisGenome [14] and PeakSeq [17] tested the significance of read enrichment against control using binomial distribution. Additionally, a novel approach was implemented by Tuteja *et al.* in their tool GLITR, in which the absolute read counts and the fold-change of a peak were mapped to a 2D space for FDR estimation [18]. These tools efficiently reduced the false positives caused by intrinsic biases and provided better estimation of FDR. The concept of control-based ChIP-seq analysis can also be extended for the comparative analysis between different cell types. Xu *et al.* developed a tool called ChIPDiff, in which an HMM was employed to identify the differential histone modification sites (DHMSs) [19].

The above-mentioned ChIP-seq analysis tools are summarized in Table **1** for the reader's comparison.

APPLICATION OF CHIP-SEQ TO TRANSCRIPTION FACTOR STUDY IN MESC

In this section, we employ Chen *et al.*'s dataset [1] in mESC to demonstrate the ChIP-seq analysis procedures, including the identification of binding sites, motif analysis, and TF co-localization analysis. We discuss on the results and show how novel biological discoveries were achieved through ChIP-seq analysis.

Dataset

The ChIP-seq dataset consists of 16 libraries, including 13 sequence-specific TFs (Nanog, Oct4, STAT3, Smad1, Sox2, Zfx, c-Myc, n-Myc, Klf4, Esrrb, Tcfcp2l1, E2f1, and CTCF), 2 transcription regulators (p300 and Suz12), and a negative control (GFP). The 15 factors in the experiment design are known to play different roles in ES-cell biology as components of the LIF and BMP signaling pathways, self-renewal regulators, and key reprogramming factors. The libraries were prepared on E14 mouse ES cells.

Identifying Binding Sites

By adopting the idea of FindPeaks, we identified the binding sites in the following four steps:

a) Each read was aligned to genome assembly using the Eland software provided by Illumina Inc. The aligned reads were extended by 200 bps toward their orientations to form virtual fragments. For each library, a ChIP-seq intensity profile was generated by projecting the virtual fragments to the reference genome (Fig. **2a**). Given a genomic location, the intensity was measured by the number of virtual fragments overlapping that location.

b) A peak-finding algorithm was applied to the intensity profile. A peak was defined as the local maxima in the profile. In detail, a genomic location with maximum intensity in its ±500 bp proximal region was selected as a candidate peak. If there were more than one local peak with the same intensity within 500bps, the one to the left in the genome assembly was chosen.

c) Monte Carlo simulation was performed by randomly extracting 200 bp fragments from the genome assembly and estimating the expected numbers of non-specific peaks with variant intensity values, where we assume a random model of uniform distribution. Given an intensity value, the FDR was estimated as the ratio of expected number of random peaks to the number of observed peaks with intensities higher or equal to that value. The cut-off threshold was set to be the minimum intensity that satisfies the criterion of FDR<0.05 or FDR<0.01, dependent on specific library.

d) To remove the falsely predicted peaks that were introduced due to the intrinsic bias of ChIP-seq, we further filtered the peaks based on the fold-change of peak intensity against the negative control library. Let $I_p(l)$ represent the binding intensity of a peak with the center at genomic locus l in a positive library, and $I_c(l)$ represent the corresponding intensity at the same locus in the negative control library, their fold-change was calculated by the ratio of $I_p(l)$ to $I_c(l)$, normalized by the sequencing depth:

$$FD(l) = \frac{I_p(l) + d_0}{I_c(l) + d_0} \times \frac{N_p}{N_c}$$

Where N_p and N_c represent the total number of aligned reads in the TF ChIP-seq library and control library, respectively. d_0 represents the *Dirichlet* prior and was set to be 1. The peaks with a fold-change below 5.0 are removed.

As the result, we identified hundreds to tens of thousands binding peaks for different libraries. Quantitative PCR (qPCR) experiments showed the predictions are highly reliable with the positive rate larger than 90%. Interestingly, the enrichment fold-change measured by qPCR validation is almost linearly correlated with the ChIP-seq intensity of the peaks (Fig. **2b**). This indicates that ChIP-seq analysis not only renders qualitative representation of binding sites, but also provides quantitative information of real binding affinity. Table **2** summarizes the information of libraries and the peak prediction.

Table 2: A summary of library information and peak prediction results from Chen *et al.*'s dataset.

Factor Name	Number of Total Uniquely Aligned Reads	Number of Predicted Peaks	FDR	Positive Rate by qPCR Validation (> 3 fold)
Nanog	8, 424, 102	10, 343	0.01	100%
Oct4	4, 911, 144	3, 761	0.01	92%
Sox2	4, 821, 446	4, 526	0.01	100%
Smad1	3, 338, 896	1, 126	0.05	96%
p300	5, 424, 357	524	0.05	90%
E2f1	8, 449, 181	20, 699	0.01	100%
Tcfcp2I1	8, 787, 961	26, 910	0.01	92%
Suz12	5, 986, 520	4, 215	0.01	100%
CTCF	3, 686, 056	39, 609	0.01	95%
Zfx	3, 844, 429	10, 338	0.01	100%
Stat3	5, 351, 116	2, 546	0.01	97%
Klf4	3, 807, 970	10, 875	0.01	95%
n-myc	4, 823, 212	7, 182	0.01	100%
c-myc	6, 637, 404	3, 422	0.01	95%
Esrrb	7, 148, 528	4, 7888	0.01	>90%
GFP (control)	3, 578, 466	-	-	-

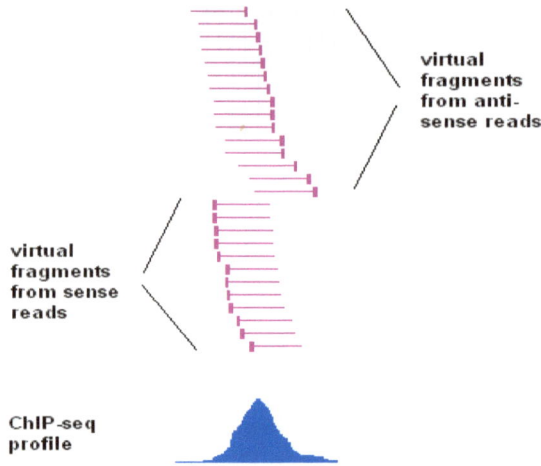

(a)

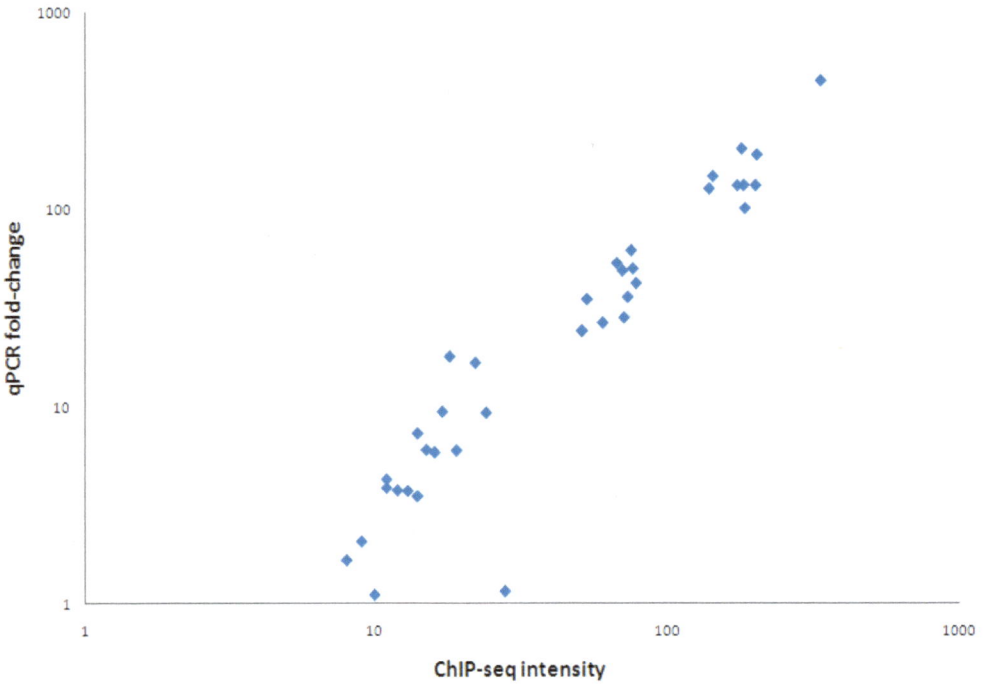

(b)

Figure 2: (a) A demonstration of virtual fragments and the ChIP-seq intensity profile. **(b)** ChIP-seq intensity is almost linearly correlated with qPCR fold-change measurement. Data retrieved from the Nanog library in Chen *et al.*'s dataset. The Pearson correlation $\rho = 0.947$.

Motif Discovery

Two of the major characteristics of ChIP-seq are the high specificity and the high-resolution representation of binding sites. Taking advantage of these characteristics, *de novo* motifs could be easily found by searching the consensus DNA sequences among the proximal regions of hundreds to tens of thousands predicted peaks. To date, motif discovery is becoming one of the routine procedures in the post-analysis of ChIP-seq data designed for TF study.

To determine the DNA-binding motifs of the TFs under study, sequences ±100 bp from the top 500 binding peaks with highest intensities were selected for each factor. Genomic repeats were masked since the redundant sequences in repeat regions will result in significant false positives in motif discovery. The program Weeder [20] was used to find overrepresented consensus sequence. Because of the high resolution in defining the binding sites offered by the high sequence depth coverage, overrepresented motifs could be uncovered from 12 of the 13 sequence-specific transcription factors (excluding E2f1) (Fig. **3a**). Among them, the Tcfcp2l1 motif has not been reported previously, and the rest eleven motif matrices are very close to the corresponding consensus sequences in literature or public motif dataset.

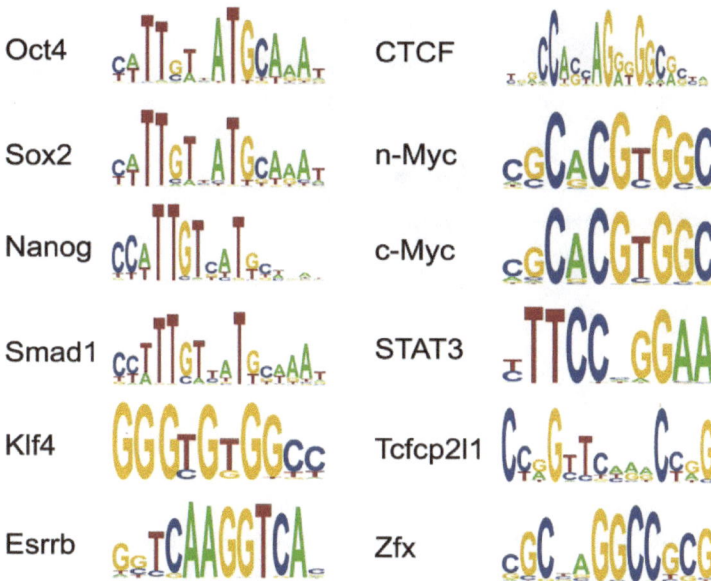

(a)

Fig. 3: cont…..

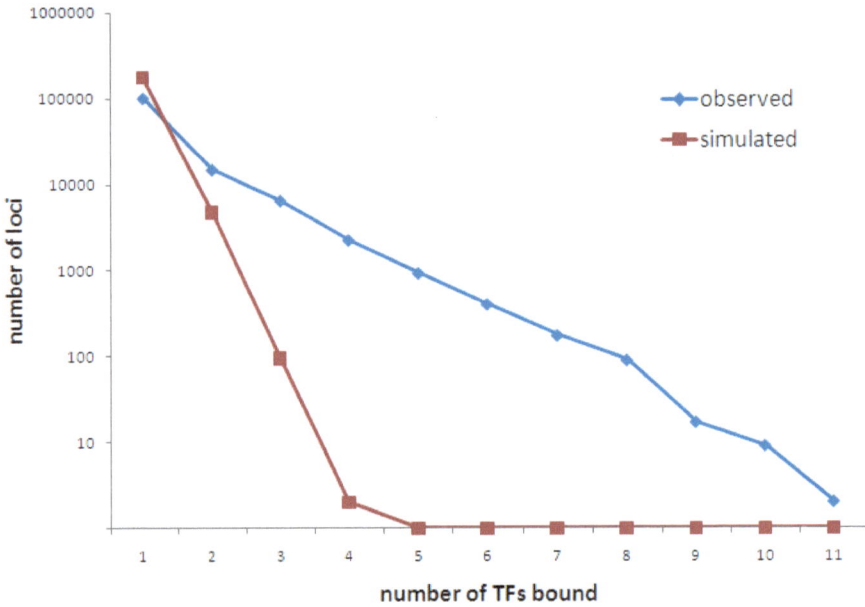

(b)

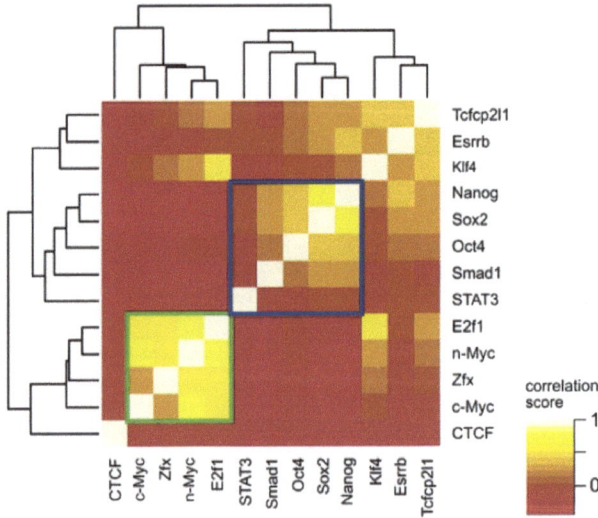

(c)

Figure 3: (a) *De novo* motif matrices discovered from ChIP-seq peaks. **(b)** Histogram of the number of TFs bound per co-bound locus. The distribution of randomly occurring co-bound loci is obtained by simulation. **(c)** Co-occurrence of TF groups within MTLs. Colors in the heat map reflect the correlation of each pair of TFs in MTLs..Images **(a)** and **(c)** in courtesy of Chen *et al.*'s article published in *Cell* [1].

Interestingly, we found the de novo motifs of the key ESC-specific regulators Nanog, Oct4, Sox2, and the signaling component Smad1, are highly similar and resemble the Oct4-Sox2 co-motif identified in our previous study [21]. This observation suggests that these factors may be co-localized to regulate mESC. To validate this hypothesis, and to find out other TFs that may involve in the co-localization, systematic computational analysis on the TF co-localization is necessary.

TF Co-Localization Analysis

The co-localization analysis started with a TFBS clustering algorithm. 189, 225 predicted binding peaks from 13 TF libraries were included as input for clustering. Consecutive binding peaks within 100 bps are iteratively grouped into co-localization loci. Each locus was assigned a co-localization score, defined as the number of TFs with binding peak (s) at that locus. To check the statistical significance of co-localization, we randomly assigned the peaks to the genome, and performed the clustering procedure on the randomized peaks. The histogram of co-localization score from predicted peaks was plotted against that from random simulation. As shown in Fig. **3b**, Loci bound by 4 or more TFs are highly significant (FDR < 0.001), and there is a total of 3583 such Loci, named Multiple TF-binding Loci (MTL). Therefore, result of the clustering method indicates that the co-localization of TFs is statistically significant at the predicted MTLs.

To further dissect the composition of the MTLs, we examined the co-occupancy of different factors found in the 3583 MTL. We rendered a TF-loci occupancy matrix, in which a Boolean value in the ith row and the jth column represents the availability of the jth TF at the ith MTL. From this matrix, we compute the Pearson correlation between each pair of columns to measure the co-occupancy of two TFs. The Pearson correlations were then used to group the highly co-occupied TFs based on hierarchical clustering (Fig. **3c**). Among the 13 TFs, Nanog, Sox2, Oct4, Smad1, and STAT3 tend to co-occur quite often, as do members of a second, distinct group comprised of n-Myc, c-Myc, E2f1, and Zfx. Strikingly, we found more than 70% of the p300 binding sites (p<1e-300) are in the MTLs with Nanog, Sox2, Oct4, Smad1 or STAT3. One the other hand, MTLs with n-Myc, c-Myc, E2f1, or Zfx are not enriched by p300 peaks, but are mostly occur in the promoter of genes.

From a perspective of biological study, our results indicate that Smad1 and STAT3 share many common target sites with Nanog, Oct4, and Sox2, and reflects a point of convergence of the two key signaling pathways (*via* Smad1 and STAT3) with the core circuitry defined by Nanog, Oct4, and Sox2. The co-localization of these factors is also associated with the co-factor p300, which is known to be one of the markers of enhancers [22]. Consequently, the MTLs bound by these factors define a number of genomic "hotspots" that regulate the gene expression in mESC.

CONCLUSION

With the advent of ultra-high-throughput sequencing technologies, ChIP-seq is becoming the main stream for the genome-wide studies of transcription factor binding sites and histone modifications. The requirement of computational methods for ChIP-seq data analysis established a new field in bioinformatics and computational biology. The ChIP-seq analysis involves the prediction of ChIP-enriched sites and the post-analysis procedures such as *de novo* motif-finding, multiple library comparison, transcription factor clustering, and gene association study. In this chapter, we reviewed the state-of-the-art computational tools for predicting ChIP-enriched sites, and employed a dataset on mESC to demonstrate the ability of some post-analysis methods. With the rapid growth of publicly available ChIP-seq dataset, one may expect that the computational analysis of ChIP-seq will play a more and more important role in the research area of transcriptional regulation in living cells.

REFERENCES

[1]   Chen X, Xu H, Yuan P, *et al.* Integration of External Signaling Pathways with the Core Transcriptional Network in Embryonic Stem Cells. Cell 2008; 133:1106-1117.
[2]   Wheeler DA, Srinivasan M, Egholm M, *et al.* The complete genome of an individual by massively parallel DNA sequencing. Nature 2008; 452:872-876.
[3]   Nagalakshmi U, Wang Z, Waern K, *et al.* The transcriptional landscape of the yeast genome defined by RNA sequencing. Science 2008; 320:1344-1349.
[4]   Cokus SJ, Feng S, Zhang X, Chen Z, *et al.* Shotgun bisulphite sequencing of the Arabidopsis genome reveals DNA methylation patterning. Nature 2008; 452:215-219.
[5]   Lieberman-Aiden E, Berkum NL, Williams L, *et al.* Comprehensive mapping of long-range interactions reveals folding principles of the human genome. Science 2009; 326:289-293.
[6]   Barski A, Cuddapah S, Cui K, *et al.* High-resolution profiling of histone methylations in the human genome. Cell 2007; 129: 823-837.

[7]   Johnson DS, Mortazavi A, Myers RM, Wold B. Genome-wide mapping of *in vivo* protein-DNA interactions. Science 2007; 316:1497-1502.

[8]   Mikkelsen TS, Ku M, Jaffe DB, *et al.* Genome-wide maps of chromatin state in pluripotent and lineage-committed cells. Nature 2007; 448:553-560.

[9]   Boyle AP, Guinney J, Crawford GE, Furey TS. F-Seq: a feature density estimator for high-throughput sequence tags. Bioinformatics, 2008; 24:2537-2538.

[10]   Robertson G, Hirst M, Bainbridge M, *et al.* Genome-wide profiles of STAT1 DNA association using chromatin immunoprecipitation and massively parallel sequencing. Nat Methods 2007; 4:651-657.

[11]   Valouev A, Johnson DS, Sundquist A, *et al.* Genome-wide analysis of transcription factor binding sites based on ChIP-Seq data. Nat Methods 2008; 5:829-834.

[12]   Jothi R, Cuddapah S, Barski A, Cui K, Zhao K. Genome-wide identification of *in vivo* protein-DNA binding sites from ChIP-Seq data. Nucleic Acids Res 2008; 36:5221-5231.

[13]   Zang C, Schones DE, Zeng C, Cui K, Zhao K, Peng W. A clustering approach for identification of enriched domains from histone modification ChIP-Seq data. Bioinformatics 2009; 25:1952-1958.

[14]   Ji H, Jiang H, Ma W, Johnson DS, Myers RM, Wong WH. An integrated software system for analyzing ChIP-chip and ChIP-seq data. Nat Biotechnol 2008; 26:1293-1300.

[15]   Zhang Y, Liu T, Meyer CA, *et al.* Model-based Analysis of ChIP-seq (MACS). Genome Biol 2008; 9:R137.

[16]   Vega VB, Cheung E, Palanismy N, Sung WK. Inherent signals in sequencing-based Chromatin-ImmunoPrecipitation control libraries. PLoS One, 2009; 4:e5241.

[17]   Rozowsky J, Euskirchen G, Auerbach RK, *et al.* PeakSeq enables systematic scoring of ChIP-seq experiments relative to controls. Nat Biotechnol 2009; 27:66-75.

[18]   Tuteja G, White P, Schug J, Kaestner KH. Extracting transcription factor targets from ChIP-Seq data. Nucleic Acids Res 2009; 37:e113.

[19]   Xu H, Wei CL, Lin F, Sung WK. Genome-wide Identification of Differential Histone Modification Sites from ChIP-seq Data. Bioinformatics 2008; 24:2344-2349.

[20]   Pavesi G, Mereghetti P, Mauri G, Pesole G. Weeder Web: discovery of transcription factor binding sites in a set of sequences from co-regulated genes. Nucleic Acids Res 2004; 32:W199-W203.

[21]   Loh YH, Wu Q, Chew JL, *et al.* The Oct4 and Nanog transcription network regulates pluripotency in mouse embryonic stem cells. Nat Genet 2006; 38:431-440.

[22]   Heintzman ND, Stuart RK, Hon G, *et al.* Distinct and predictive chromatin signatures of transcriptional promoters and enhancers in the human genome. Nat Genet, 2007; 39:311-318.

CHAPTER 13

The FunGenES Database: A Reference and Discovery Tool for Embryonic Stem Cells and their Derivatives

Antonis K. Hatzopoulos[*]

Vanderbilt University, TN, USA

Abstract: The "Functional Genomics in Embryonic Stem Cells" (FunGenES) database is based on gene expression profiling data obtained in selected pluripotent mouse Embryonic Stem (ES) cell lines using Affymetrix Mouse 430 v.2 arrays. The interactive FunGenES database allows users to derive gene expression profiles for every transcript that is included in the microarrays, or perform a series of gene association studies to search for groups of co-expressed and thus possibly co-regulated genes during ES cell growth and differentiation. It also includes advanced annotation tools and numerous connections to external reference tools and databases, linking gene expression patterns in stem cells to vital information about the role of corresponding genes in embryonic development or adult homeostasis and disease. In this chapter, I have used specific gene query examples to highlight the potential of the FunGenES database as a reference and discovery tool to study the biology of stem cells.

Keywords: Functional genomics, Mouse embryonic stem cells, Microarray analysis, FunGenES Database, Gene expression profiling, Bioinformatics, Gene clustering, Time waves, Expression waves, Pathway animations, Multi-Experiment Matrix, g:Profiler.

INTRODUCTION

Embryonic stem (ES) cells have practically unlimited self-renewal capacity and the potential to differentiate into a wide variety of cell types including cardiac, endothelial, hematopoietic, neuronal, or insulin-producing β-cells, thus showing great promise for organ repair [1-4]. They are also powerful experimental tools in the laboratory for drug discovery or studies on human development, ageing and diseases such as cancer [5, 6].

Although the rich differentiation potential of ES cells offers an ample cellular

*Address correspondence to A.K. Hatzopoulos:** Department of Medicine, Division of Cardiovascular Medicine, Vanderbilt Center for Stem Cell Biology, Vanderbilt University, MRB IV - P425C, 2213 Garland Avenue, Nashville, TN 37232-6300, USA; Tel: (615) 936 5529; Fax: (615) 936 4079; E-mail: antonis.hatzopoulos@vanderbilt.edu

source for experimental and clinical studies, it also presents several practical problems [4, 7]. For example, the co-differentiation of multiple cell types creates challenges for cellular therapy that mostly requires monotypic cultures of specialized cells. Furthermore, the low yield of particular cell types necessitates elaborate manipulations to enrich for cells with desired phenotypes. For this reason, understanding the programs controlling self-renewal and differentiation of stem cells may lead to novel approaches to unlock their regenerative potential. In this arena, mouse ES cells offer an accessible and relevant model system because they grow and differentiate robustly and reproducibly, recapitulate events of early embryonic development, maintain a stable phenotype over many passages and are easy to genetically engineer [4, 8, 9].

In recent years, a number of genome-wide gene expression profiling approaches provided a wealth of information on the genetic make up of mouse and human ES cells [10-18]. To fully profit from these resources, development of innovative bioinformatics tools is necessary to explore this knowledge and mine available data. In this chapter, I will describe the conceptual design and main features of the FunGenES database that was built on gene expression data obtained in mouse ES cells and may serve as template for the organization of future genomic resources.

METHODS

The History of the FunGenES Consortium

The "Functional Genomics in Embryonic Stem Cells" Consortium (acronym FunGenES; http://www.fungenes.org) was funded by the European Framework Programme 6 from 2003 to 2007. The consortium included 20 research groups, which analyzed gene expression patterns in ES cells before and after differentiation under a large number of diverse conditions. Data collection took place in coordinated fashion by streamlining experimental protocols among partners and focusing on a small number of independently-derived ES cell lines, namely the germline-competent CGR8, E14TG2a and R1 cells. Most experiments were performed in at least six separate biological replicates and the total number of conditions included in the original analysis was 67. RNA samples were prepared with the same techniques and analyzed in a central facility using 258 Affymetrix Mouse 430 v.2 arrays. Detailed descriptions of the individual experimental settings have been previously published [18-25].

Main Objectives of the FunGenES Database

The coordinated collection of gene expression data facilitated their organization in an interactive database with a number of specific objectives. The first was to provide easy access to the bulk of the FunGenES data both to Consortium partners and the scientific community at large. For this reason, database development took place in a close, step-by-step, cooperation between biologists and bioinformaticians. The second was to arrange expression profiles based on two fundamental parameters, *i.e.,* timing of expression during ES cell differentiation and gene function. Third, the database was designed as a reference tool for the expression pattern of any gene or group of genes in ES cells and their derivatives. This way data obtained in wet labs can be quickly validated *in silico*. Alternatively, before dedicating lab resources to a particular group of genes, their association with a specific differentiation process can be first evaluated using the FunGenES database.

One advantage of large-scale expression studies using RNA generated in the same system under an array of different experimental conditions is that gene clustering can piece together molecular and cellular pathways with critical roles in the biological processes under investigation. In the case of pluripotent stem cells, an extra tier of information comes from gene groups associated with specific developmental or differentiation mechanisms. Therefore, the fourth goal of the database design was to devise tools to search for transcripts that behave the same way, *i.e.,* are co-induced or co-suppressed during ES cell growth and differentiation under a number of experimental conditions. It is likely that genes linked this way are components of the same molecular, cellular or developmental pathways, or targets of common regulatory mechanisms, or both [26-28]. Because various clusters contain together well and poorly characterized co-regulated genes, the latter may participate in the same biological networks as the known genes in the same cluster. This information thus provides a starting point to discover the function of new genes that have not been thoroughly studied before, or discover new functions of genes previously analyzed under different contexts.

The fifth objective was to depict gene-clustering results in interactive visual representations to enhance discovery of new mechanisms. Therefore, effort was put into creating tools such as the Expression Waves which assembles genes with characteristic expression profiles during ES cell differentiation in the same window as genes that respond in the opposite way and Pathway Animations that illustrate dynamic changes in the components of individual signaling and metabolic pathways viewed in time-related manner.

Sixth, to further enhance the discovery tools, the database was linked to external resources including the NCBI Entrez search engine (http://www.ncbi.nlm.nih.gov /sites/gquery), iHOP (http://www.ihop-net.org), Ensembl (www.ensembl.org), Pubgene (http://www.pubgene.org) and String (http://string-db.org). There are also links to the Amazonia! (http://amazonia. transcriptome.eu), Hematopoietic Fingerprints (http://franklin.imgen.bcm.tmc.edu/loligag) and SCGAP Urologic Epithelial Stem lls Project (http://scgap.systemsbiology.net) databases. In this way, the expression patterns of genes of interest during ES cell differentiation can be compared to their corresponding profiles in adult stem cells and adult tissues. Even more powerful, is the new Multi-Experiment Matrix (MEM) feature developed by our Bioinformatics team which enables users to search for transcripts that follow similar patterns as a specific gene of interest within the broad collection of publically available gene expression profiling data [29].

Finally, to enhance the discovery of new mechanisms, the g:Profiler tool provides functional annotation to assess the biological classification of transcripts with specific expression patterns   [30]. Gene assignments in g:Profiler include GO categories [31], KEGG [32] and Reactome pathways [33], miRBase microRNA information   [34], and TRANSFAC motifs   [35]. In addition to functional annotations, g:Profiler provides tools to sort different gene identifiers and find orthologs from other organisms.

Features of the FunGenES Database

The six main components of the FunGenES database, each accessed by a window on the home page (http://biit.cs.ut.ee/fungenes), include: a) Global Clusters of genes that are co-expressed under all experimental conditions used for data collection; b) Time Series of gene expression profiles during normal ES cell differentiation of all the probe sets (*i.e.,* genes or transcripts, ESTs, unknown sequences) in the Affymetrix Mouse 430 v.2 arrays; c) Specific Gene Classes that are gene expression time series of transcriptional factors and ESTs; d) Expression Waves of genes with characteristic profiles during ES cell differentiation juxtaposed to genes that follow opposite patterns; e) Pathway Animations of dynamic changes in signaling and metabolic pathways during ES cell differentiation [36]; and, f) the Study your Gene (s) of Interest search engine to query the expression pattern of any transcript (s) during ES cell differentiation.

The computational methods employed to analyze the Affymetrix data and perform unsupervised hierarchical clustering, as well as the development of FunGenES specific algorithms and bioinformatics tools have been described in previous publications [18, 29, 30, 36]. In this chapter, I will draw on specific examples to showcase the potential of the FunGenES database as a reference and discovery tool. For this purpose, I will use the probe sets for the genes encoding the morphogen Wnt3 protein and the transcription factor GATA-1 as entry points to explore database functions.

Wnt3 and *Gata1* Gene Association Studies Using the Fungenes Database: The Clusters

The **Global Clusters** window provides an opportunity to test whether a gene belongs to one of the 115 groups generated through unsupervised hierarchical clustering of the entire collection of the FunGenES expression data [18]. In many instances, such groups consist of genes that have been associated with particular developmental processes such as neurogenesis (Cluster 4) and hemopoiesis (Cluster 20), or cellular functions such as extracellular matrix production (Cluster 9), RNA biosynthesis (Clusters 24, 28) and calcium metabolism (Cluster 31).

Wnt3 is expressed around the time that gastrulation begins in the embryo and belongs to a group of genes that performs a unique function conserved from *Drosophila* to man, controlling morphogenetic movements to set up the original body plan [37, 38]. Consistent with its biological roles, the *Wnt3* probe set belongs to cluster 30, which contains genes that are transiently expressed at day 3 of ES cell differentiation prior to induction of genes involved in mesoderm formation (Cluster 15), neurogenesis (Cluster 4), hemopoiesis (Cluster 20) and cardiogenesis (Cluster 12). Cluster 30 also includes other genes that are known to have critical roles at this embryonic stage [39], such as *Goosecoid (Gsc)*,

Cerberus like 1 homolog (Cer1), *Mesp1*, *Mixl1*, *mEomes* and *Even-skipped 1 (Evx1*; Fig. **1**).

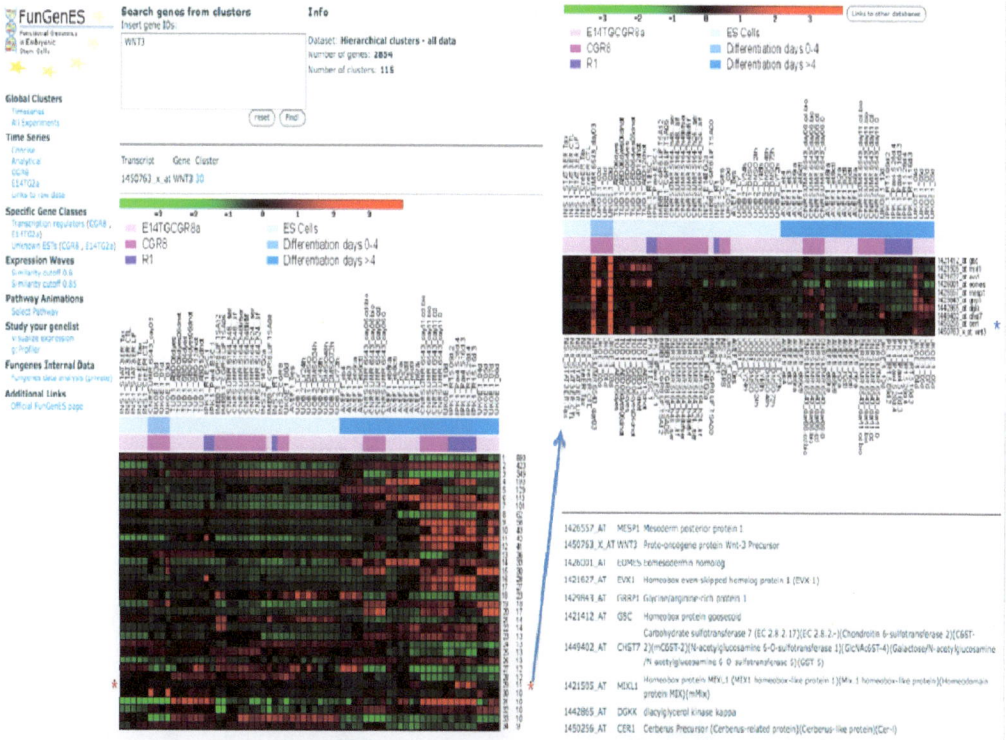

Figure 1: Search results using the FunGenES database Global Clusters with *Wnt3* as query reveal that the *Wnt3* probe set belongs to Cluster 30 (left panel, marked by red asterisks). Cluster 30 contains transiently expressed genes linked to gastrulation, body axis formation and generation of the three germ layers of endoderm, mesoderm and ectoderm. This assignment is consistent with the function of Wnt3 during embryonic development. The gene content and the corresponding expression patterns (or heat maps) of Cluster 30 members (right panel) are retrieved by clicking on the Cluster 30 heat map. *Wnt3* is marked with a blue asterisk.

Gata-1 is a key transcriptional regulator of blood development and expression of erythrocyte-specific genes [40]. In a corresponding manner, the *Gata1* probe set belongs to Cluster 20, which as mentioned above, is linked to hemopoiesis. Besides *Gata1*, this Cluster contains regulatory and structural genes encoding proteins important for blood development (*Nfe2*, *Klf1*) and erythrocytic functions such as hemoglobins (*Hba-x*, *Hbb-b1*, *Hbb-bh1*, *Hbb-b2*, *Hbb-a1*, *Hbb-a2*) and *Glycophorin A* (*Gypa*, not shown).

Besides the Global Clusters, which are derived from the entire data pool, the Time Series tool has been designed to visualize gene expression profiles during ES cell differentiation using a subset of samples representing normal growth and differentiation conditions, *i.e.,* without addition of exogenous agents except standard serum. Therefore, the Time Series assembly offers a comprehensive collection of genes with specific expression patterns during ES cell differentiation and, by extension, early embryonic development. Time Series have been organized in 50 Concise and 200 Analytical clusters, the latter having more detailed time resolution of gene expression patterns.

Inspection of the 50 Concise clusters, reveals that Time Series Concise Cluster 3, which represents transiently induced genes around day 3 of differentiation, contains the same ten transcripts as Global Cluster 30 including *Wnt3*. However, Time Series Cluster 3 is much larger (59 transcripts) comprising in addition genes such as *T-brachyury (T)*, *Axin2*, *Mesp2*, *Fgf8*, *Wnt8a*, *Sp5*, *Sp8*, *Follistatin*, *Mixl1* and *Lim1*, which are also known to take part in the same developmental phase of embryogenesis as *Wnt3* (Fig. **2**).

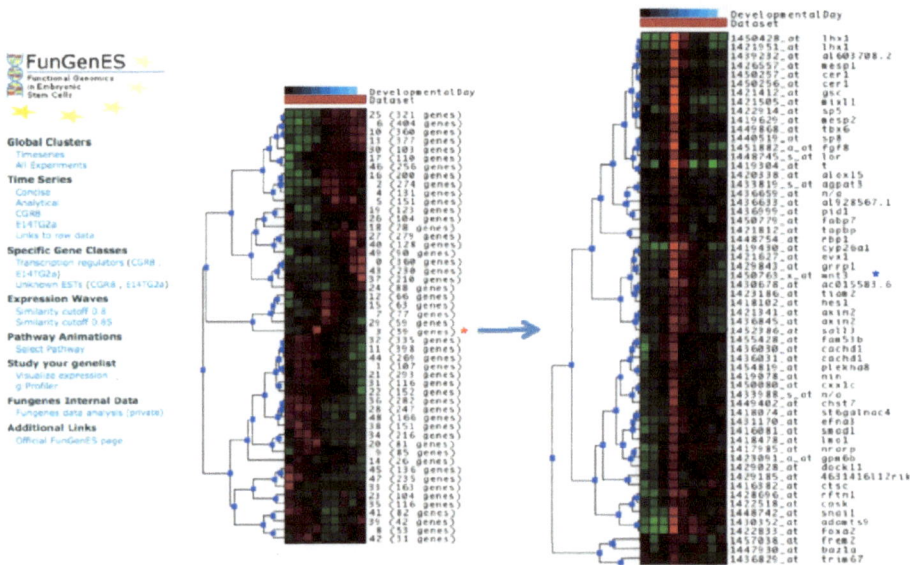

Figure 2: Captions of the FunGenES Time Series 50 Concise Clusters show that Concise Cluster 3 (marked by a red asterisk) follows the *Wnt3* expression pattern, *i.e.,* contains genes transiently induced at day 3 of ES cell differentiation. Time Series 3 contents and gene expression profiles including *Wnt3* (right panel, blue asterisk) are retrieved by clicking on the heatmap of Cluster 3.

The *Gata1* transcript is in Time Series Analytical Cluster 137 that contains a group of 49 transcripts enriched in blood-specific genes encoding the Rhesus blood group-associated A glycoprotein gene (*RHAG*), erythrocyte membrane protein band 4.2 (Epb4.2) and *Hemogen* (*HEMGN*), which is also a known GATA-1 target [41]. However, it should be noted that Cluster 137 also contains a number of endothelial-specific genes (*e.g., Tek, Tie1, Sox18*), a cell-type that co-differentiates with erythroblasts in the blood islands of the yolk sac during early development. Therefore, time-of-expression clusters may often contain genes associated with different processes that proceed in parallel fashion during embryonic development and ES cell differentiation. Finally, *Gata1* is part of Specific Gene Classes Cluster 9 (24 transcripts) of transcriptional regulators. Besides *Gata1*, this cluster also contains other transcriptional regulators in various blood cell types such as *Nfe2, Klf1, Erg, Ets2, Ikfz1, Lyl1, etc.* A small number of transcriptional factors linked to early cardiovascular development that takes place in the same time window of ES cell differentiation (*Sox18, Gata5*) also associate with this group.

Exploring Gene Functions with the FunGenES Database: Study Your Genes of Interest

The analysis described above depends on the query gene being present in one of the Global Clusters as a first step to classify its expression pattern and identify co-expressed genes, serving as a starting point for further functional associations. An alternative way is to go directly to the search engine named Study your gene (s) of Interest and enter a standard abbreviated gene name, an Affymetrix probe set ID, or any other identifier supported by the Ensembl database. The search engine will configure the expression pattern of the gene during ES cell differentiation and provide access to a wide array of external resources to collect information related to this particular gene (or genes).

For example, after initiating a search using *Wnt3*, a new window offers a number of options including links to various stem cell (Stem cell links) and genomic (General resource links) databases, as well as selections to view expression in FunGenES expression datasets and obtain functional annotation or perform additional queries (Fig. **3**). Selecting View expression on FunGenES matrices opens a new window allowing users to visualize expression of genes in the two

CGR8 and E14TG2a ES cell lines either separately or in both combined (Time series; Fig. **3**).

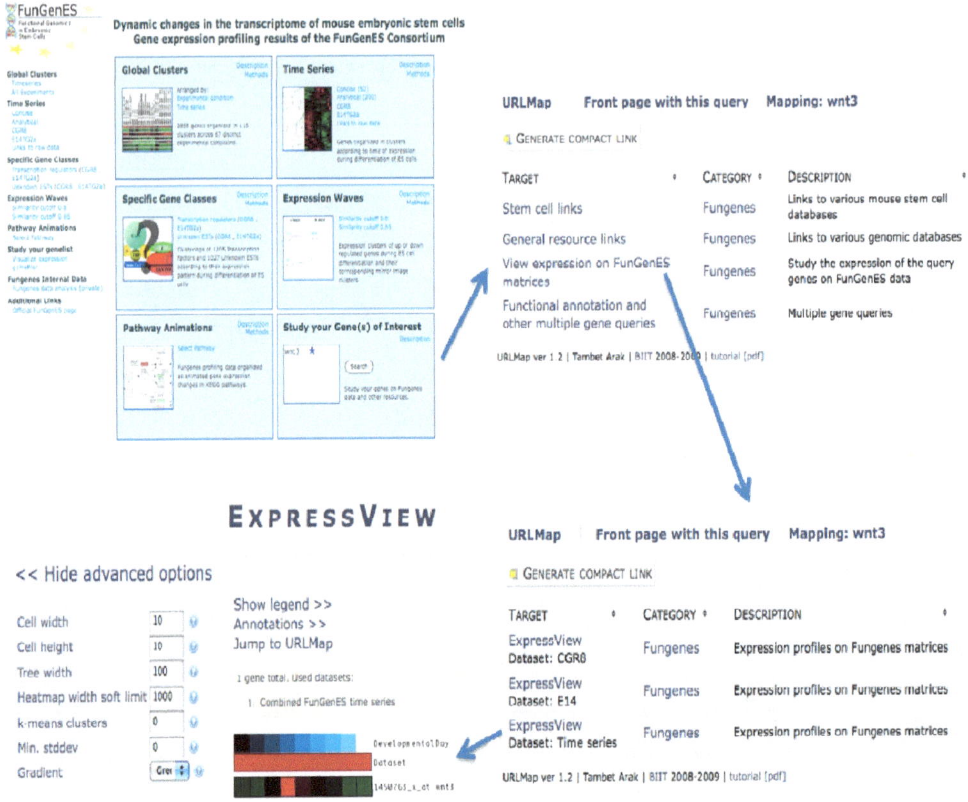

Figure 3: Search options using the FunGenES Study your Gene (s) of Interest engine with *Wnt3* as query. Entering the standard *Wnt3* abbreviated name (blue asterisk) and initiating Search, opens a new window providing the option to obtain expression of *Wnt3* during ES cell differentiation in CGR8, E14TG2a (E14) or both as Time series (View expression on Fungenes matrices). Selecting Time series generates the heatmap of *Wnt3* (under ExpressView) showing expression at Developmental Day 3. The "Show Legend" button provides descriptions of color codes for Developmental Day, heatmap range and dataset used. Additional options (bottom left) offer a range of heatmap color selections, display sizes, *etc.* as explained when engaging the adjacent "?" buttons.

The Stem cell links offers the opportunity to view expression of genes in different stem cell types and adult tissues with links to Amazonia! [42], Hematopoietic Fingerprints [43] and SCGAP Urologic Epithelial Stem Cells Project [44] databases. An example from Amazonia! is showing high *Wnt3* expression in adult skin tissue samples (Fig. **4**). The Functional annotation and other multiple gene queries leads to a number of resources such as relevant literature information about the function of *Wnt3*, or to tools recognizing *Wnt3* gene IDs and orthologs in other organisms.

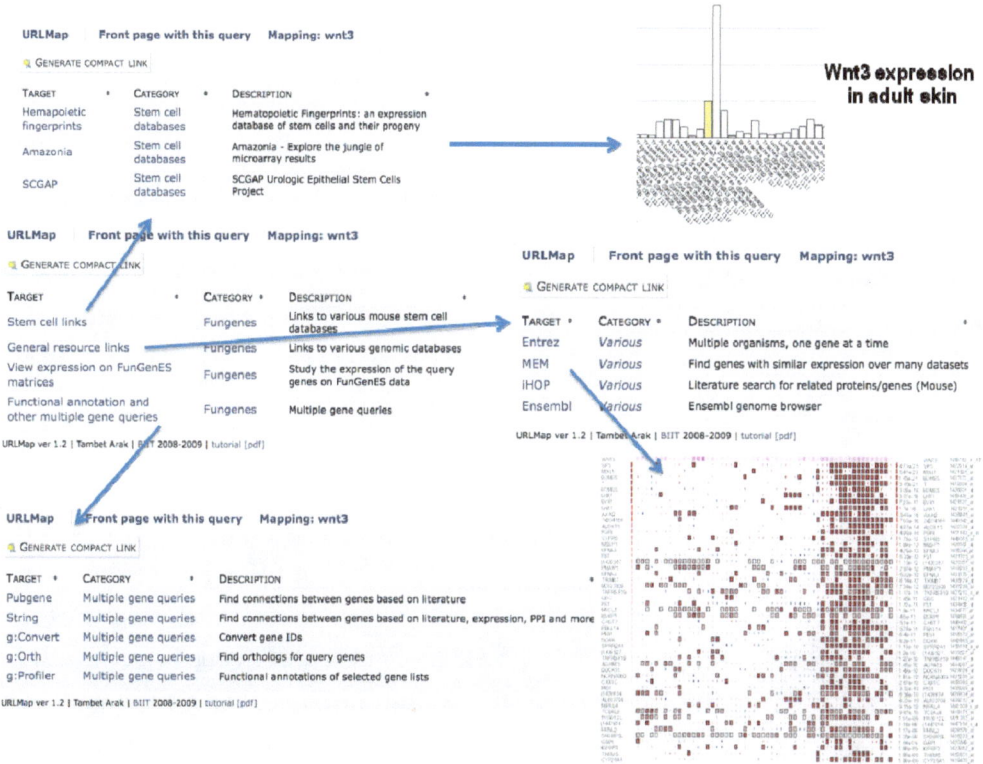

Figure 4: Additional search options using the FunGenES Study your Gene (s) of Interest engine with *Wnt3* as query. Besides View expression on Fungenes matrices described in Fig. **(3)**, Stem cell links access gene expression in the Amazonia!, Hematopoietic Fingerprints and SCGAP Urologic Epithelial Stem Cells Project databases. An example from Amazonia! showing high *Wnt3* expression in adult skin is depicted (top right). The Functional annotation and other multiple gene queries leads to information pertinent to *Wnt3* (PubMed, String), or to various annotations tools (g:Convert, g:Orth, and g:Profiler). The General resource links provides access to additional databases (Entrez, iHOP, Ensembl), whereas the Multi-Experiment Matrix (MEM) search engine identifies genes with similar expression patterns as *Wnt3* within publically available microarray data. MEM search results for *Wnt3* are included (bottom right).

The General resource links provides access to a variety of genetic databases (Entrez, iHOP, Ensembl) and the new Multi-Experiment Matrix (MEM) search engine [29]. The MEM tool is able to search for genes with similar expression patterns as *Wnt3* within the large collection of publically available expression profiling data and connect this to the actual samples and experimental conditions used by previous investigations. Specifically, the outcome using *Wnt3* brings a compilation of genes that includes among others *Gsc, Mesp1, Mixl1, mEomes, Evx1, T-brachyury, Axin2, Mesp2, Fgf8, Wnt8a, Sp5, Follistatin,* and *Lim1* (Fig. **4**). In similar manner, search with *Gata1* recovers *Nfe2, Klf1, Gypa, HEMGN* and *RHAG* (not shown).

The fact that the gene group assembled by MEM using the *Wnt3* and *Gata1* genes as baits overlaps extensively with the content of the FunGenES Global and Time Series Clusters, and because most of these genes have been implicated in the same biological processes during development or in adult tissues, provides further support that the majority of the genes within FunGenES Clusters are functionally related.

Expression Waves and Pathway Animations

Two additional tools to view co-regulated genes have been included in the FunGenES database. The first, called Expression Waves, was generated by drawing a series of specific, predetermined expression profiles during ES cell differentiation. Next, transcripts were assigned to the graph or expression wave that most closely follows their expression profile. A row of graphs representing genes that are first induced at day 3 ES cell differentiation and follow different expression patterns thereafter is shown in Fig. (**5**). The first graph to the left represents genes that are transiently expressed at day 3. The names of the genes belonging to the corresponding Expression Wave appear below by clicking on the graph. In the above example, transiently induced genes at day 3 include as expected *Wnt3, Gsc, Cer1, Mesp1, Mixl1, T-brachyury, Mesp2, Fgf8* and *Sp5*. A window link to the gene group expressed in the opposite manner is available on the same page for side-by-side comparisons to reveal potentially co-induced and co-suppressed genes. Interestingly, a single gene, Collagen 18 α1, which is a key component of basal membranes surrounding all cells and tissues, is downregulated during this phase. It is possible that suppression of Collagen 18 α1 is required to allow cell migration and morphogenetic movements during gastrulation.

An additional tool gives investigators the opportunity to visualize sequential stages or waves of gene expression during ES cell differentiation in a comprehensive manner by selecting the corresponding Waves using the boxes above the graphs and then applying the "Merge" button.

A related tool to illustrate sequential gene expression changes is the Pathway Animations that depicts dynamic changes in specific genetic, signaling or metabolic pathways (based on KEGG annotations) viewed in time animations [32, 36]. The resource offers a set of functions that allow users to reanimate the graphs by selecting specific time points and/or subsets of pathway components. A capture of the wnt pathway at day 3 of ES cell differentiation reveals that the stage is characterized by high levels of *Wnt3*, *Cer1* and *Axin2* in accordance to the results obtained using the Global Clusters and Time Series (Fig. **5**). Furthermore, the tool illustrates other highly expressed pathway components at this particular stage such as the transcription factor Sox17. Although Sox17 does not cluster with *Wnt3*, *Cer1* and *Axin2,* (possibly because is also expressed later in endodermal cells), its high expression levels suggests that is one of the downstream effectors of Wnt signaling at this differentiation stage.

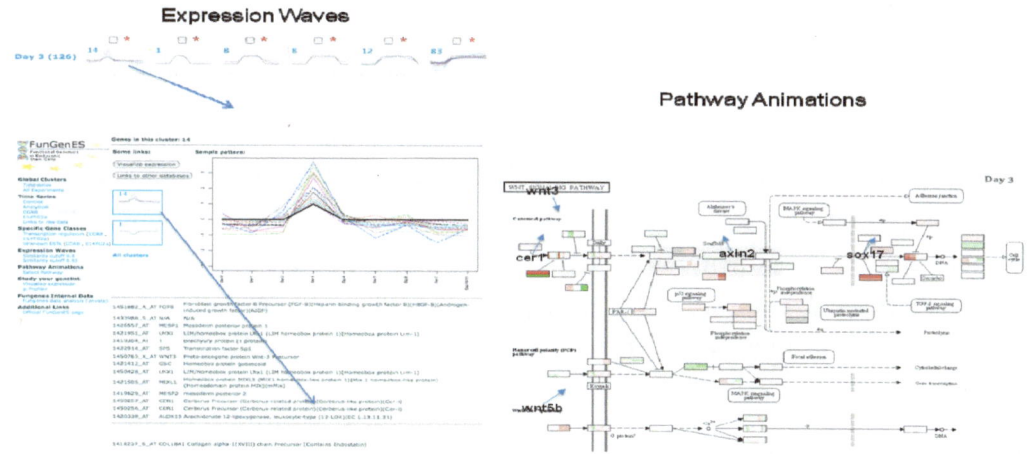

Figure 5: Display windows using the Expression Waves (left) and Pathway Animations (right) tools. Top left illustrates expression profiles of genes that are induced at day 3. The first wave, containing 14 transcripts, is only transiently active for one day. Genes the expression of which follows this pattern are shown below after clicking on the graph. The new window also includes the graph of genes that are suppressed during this wave. Marking the boxes above the graphs (red asterisks) allows users to visualize together the gene content of the selected waves. The right panel

depicts a single frame of the wnt signaling pathway animation at day 3 of ES cell differentiation. Each box represents the expression profile of a single gene or a family of related genes that is disclosed by placing the cursor over the box. Red color signifies expression, green absence of expression. Vertical colored lines represent independent genes. Horizontal lines denote the expression of different probe sets for the same gene included in the microarray. For simplicity, only the names of few selected genes have been included with the corresponding boxes marked by arrows.

The specific examples described above highlight some of the FunGenES uses, but there are more functions and tools embedded in every window. Feel free to explore the database and contact us with questions, comments or suggestions.

CONCLUDING REMARKS

High throughput genomic approaches, sophisticated bioinformatics analyses and functional assays have begun to piece together the networks regulating the growth and differentiation of ES cells [10, 45-48]. These studies indicate that the mechanisms regulating pluripotency and early differentiation in ES cells is more complex than previously thought [49]. Moreover, recent techniques to derive ES cell-like cells from adult tissues called induced Pluripotent Stem (iPS) cells and the heterogeneity of the derived cell lines further underscore the need for systematic analysis of the ES and iPS cell transcriptomes to establish universal standards of "stemness" [50-53]. Providing tools to perform gene association studies, access to external resources and databases, search engines linking gene expression profiles in ES cells to those in the vast public collection of microarray data, and advanced annotations tools, the FunGenES database (and comparable resources [15, 54]) may serve as a template for future ES and iPS databases, become a useful reference tool for scientists interested in the biology of stem cells and lead to new discoveries.

CONFLICT OF INTEREST

None declared.

ACKNOWLEDGEMENTS

The FunGenES Integrated Project was funded by the European Framework Programme 6. Jürgen Hescheler was the Coordinator of the FunGenES consortium, Herbert Schulz performed the global data clustering analysis and Jack Vilo managed the bioinformatics team that built the public FunGenES Database. This work was also supported by NIH grants HL083958 and HL100398 to AKH.

REFERENCES

[1]   Evans MJ, Kaufman MH. Establishment in culture of pluripotential cells from mouse embryos. Nature 1981; 292: 154-156.

[2]   Martin GR. Isolation of a pluripotent cell line from early mouse embryos cultured in medium conditioned by teratocarcinoma stem cells. Proc Natl Acad Sci USA 1981; 78: 7634-7638.

[3]   Thomson JA, Itskovitz-Eldor J, Shapiro SS, *et al.,* Embryonic stem cell lines derived from human blastocysts. Science 1998; 282: 1145-1147.

[4]   Murry CE, Keller G. Differentiation of Embryonic Stem Cells to clinically relevant populations: Lessons from embryonic development. Cell 2008; 132: 661-680.

[5]   Singec I, Jandial R, Crain A, Nikkhah G, Snyder EY. The leading edge of stem cell therapeutics. Annu Rev Med 2007; 58:313-328.

[6]   Rando TA. Stem cells, ageing and the quest for immortality. Nature 2006; 441: 1080-1086.

[7]   Boudoulas KD, Hatzopoulos AK. Cardiac repair and regeneration: the Rubik's cube of cell therapy for heart disease. Dis Model Mech 2009; 2: 344-358.

[8]   Doetschman TC, Eistetter H, Katz M, Schmidt W, Kemler R. The *in vitro* development of blastocyst-derived embryonic stem cell lines: formation of visceral yolk sac, blood islands and myocardium. J Embryol Exp Morphol 1985; 87: 27-45.

[9]   Thomas KR, Capecchi MR. Site-directed mutagenesis by gene targeting in mouse embryo-derived stem cells. Cell 1987; 51: 503-512.

[10]   Ivanova NB, Dimos JT, Schaniel C, Hackney JA, Moore KA, Lemischka IR. A stem cell molecular signature. Science 2002; 298: 601-604.

[11]   Ramalho-Santos M, Yoon S, Matsuzaki Y, Mulligan RC, Melton DA. "Stemness": Transcriptional profiling of embryonic and adult stem cells. Science 2002; 298: 597-600.

[12]   Sato N, Sanjuan IM, Heke M, Uchida M, Naef F, Brivanlou AH. Molecular signature of human embryonic stem cells and its comparison with the mouse. Dev Biol 2003; 260: 404-413.

[13]   Sekkai D, Gruel G, Herry M, *et al.,* Microarray analysis of LIF/Stat3 transcriptional targets in embryonic stem cells. Stem Cells 2005; 23: 1634-1642.

[14]   Ivanova N, Dobrin R, Lu R, Kotenko I, *et al.,* Dissecting self-renewal in stem cells with RNA interference. Nature 2006; 442: 533-538.

[15]   Assou S, Le Carrour T, Tondeur S, *et al.,* A meta-analysis of human embryonic stem cells transcriptome integrated into a web-based expression atlas. Stem Cells 2007; 25: 961-973.

[16]   Cinelli P, Casanova E, Uhlig S, *et al.,* Expression profiling in transgenic FVB/N embryonic stem cells overexpressing STAT3. BMC Development Biol 2008; 8: 57.

[17]   Müller FJ, Laurent LC, Kostka D, *et al.,* Regulatory networks define phenotypic classes of human stem cell lines. Nature 2008; 455: 401-405.

[18]   Schulz H, Kolde R, Adler P, *et al.,* The FunGenES database: a genomics resource for mouse Embryonic Stem cell differentiation. PLoS One 2009, 4: e6804.

[19] Doss MX, Winkler J, Chen S, *et al.*, Global transcriptome analysis of murine embryonic stem cell-derived cardiomyocytes. Genome Biol 2007; 8: R56.

[20] Doss MX, Chen S, Winkler J, *et al.*, Transcriptomic and phenotypic analysis of murine embryonic stem cell derived BMP2+ lineage cells: an insight into mesodermal patterning. Genome Biol 8 2007: R184.

[21] Karantzali E, Schulz H, Hummel O, Huebner N, Hatzopoulos AK, Kretsovali A. Histone deacetylase inhibition accelerates the early events of stem cell differentiation: transcriptomic and epigenetic analysis. Genome Biol 2008; 9: R65.

[22] Potta SP, Liang H, Pfannkuche K, *et al.*, Functional characterization and transcriptome analysis of embryonic stem cell-derived contractile smooth muscle cells. Hypertension 2009; 53: 196-204.

[23] Mariappan D, Winkler J, Chen S, Schulz H, Hescheler J, Sachinidis A. Transcriptional profiling of CD31 (+) cells isolated from murine embryonic stem cells. Genes Cells 2009; 14: 243-260.

[24] Trouillas M, Saucourt C, Guillotin B, *et al.*, Three LIF-dependent signatures and gene clusters with atypical expression profiles, identified by transcriptome studies in mouse ES cells and early derivatives. BMC Genom 2009; 10: 73.

[25] Rolletschek A, Schroeder IS, Schulz H, Hummel O, Huebner N, Wobus AM. Characterization of mouse embryonic stem cell differentiation into the pancreatic lineage *in vitro* by transcriptional profiling, quantitative RT-PCR and immunocytochemistry. Int J Dev Biol 2009; 54: 41-54.

[26] Eisen MB, Spellman PT, Brown PO, Botstein D. Cluster analysis and display of genome-wide expression patterns. Proc Natl Acad Sci USA 1998; 95: 14863-14868.

[27] Brown MP, Grundy WN, Lin D, *et al.*, Knowledge-based analysis of microarray gene expression data by using support vector machines. Proc Natl Acad Sci USA 2000; 97: 262-267.

[28] Wu LF, Hughes TR, Davierwala AP, Robinson MD, Stoughton R, Altschuler SJ. Large-scale prediction of Saccharomyces cerevisiae gene function using overlapping transcriptional clusters. Nat Genet 2002; 31: 255-265.

[29] Adler P, Kolde R, Kull M, Tkachenko A, Peterson H, Reimand J, Vilo J. Mining for coexpression across hundreds of datasets using novel rank aggregation and visualization methods. Genome Biol 2009; 10: R139.

[30] Reimand J, Kull M, Peterson H, Hansen J, Vilo J. g:Profiler--a web-based toolset for functional profiling of gene lists from large-scale experiments. Nucleic Acids Res 2007; 35: W193-W200.

[31] Ashburner M, Ball CA, Blake JA, *et al.*, Gene Ontology: tool for the unification of biology. Nat Genet 2000; 25: 25-29.

[32] Kanehisa M, Goto S. KEGG: kyoto encyclopedia of genes and genomes. Nucleic Acids Res 2000; 28: 27-30.

[33] Vastrik I, D'Eustachio P, Schmidt E, *et al.*, Reactome: a knowledge base of biologic pathways and processes. Genome Biol 2007; 8: R39.

[34] Griffiths-Jones S, Grocock RJ, van Dongen S, Bateman A, Enright AJ. miRBase: microRNA sequences, targets and gene nomenclature. Nucleic Acids Res 2006; 34: D140-D144.

[35] Wingender E, Dietze P, Karas H, Knuppel R. TRANSFAC: a database on transcription factors and their DNA binding sites. Nucleic Acids Res 1996; 24: 238-241.

[36]   Adler P, Reimand J, Jänes J, Kolde R, Peterson H, Vilo J. KEGGanim: pathway animations for high-throughput data. Bioinformatics 2008; 24: 588-590.

[37]   Barrow JR, Howell WD, Rule M, *et al.,* Wnt3 signaling in the epiblast is required for proper orientation of the anteroposterior axis. Dev Biol 2007; 312: 312-320.

[38]   Takaoka K, Yamamoto M, Hamada H. Origin of body axes in the mouse embryo. Curr Opin Genet Dev 2007; 17: 344-350.

[39]   Tam PP, Loebel DA. Gene function in mouse embryogenesis: get set for gastrulation. Nat Rev Genet 2007; 8: 368-381.

[40]   Migliaccio AR, Rana RA, Vannucchi AM, Manzoli FA. Role of GATA-1 in normal and neoplastic hemopoiesis. Ann N Y Acad Sci 2005; 1044: 142-158.

[41]   Yang LV, Wan J, Ge Y, *et al.,* The GATA site-dependent hemogen promoter is transcriptionally regulated by GATA1 in hematopoietic and leukemia cells. Leukemia 2006; 20: 417–425

[42]   Le Carrour T, Assou S, Tondeur S, *et al.,* Amazonia!: An online resource to google and visualize public human whole genome expression data. The Open Bioinformatics J 2010; 4: 5-10.

[43]   Chambers SM, Boles NC, Lin KY, *et al.,* Hematopoietic fingerprints: an expression database of stem cells and their progeny. Cell Stem Cell 2007; 1: 578-591.

[44]   Oudes AJ, Campbell DS, Sorensen CM, Walashek LS, True LD, Liu AY. Transcriptomes of human prostate cells. BMC Genomics 2006; 7: 92.

[45]   Boyer LA, Lee TI, Cole MF, *et al.,* Core transcriptional regulatory circuitry in human embryonic stem cells. Cell 2005; 122: 947-956.

[46]   Loh YH, Wu Q, Chew JL, *et al.,* The Oct4 and Nanog transcription network regulates pluripotency in mouse embryonic stem cells. Nat Genet 2006; 38: 431-440.

[47]   Zhou Q, Chipperfield H, Melton DA, Wong WH. A gene regulatory network in mouse embryonic stem cells. Proc Natl Acad Sci USA 2007; 104: 16438-16443.

[48]   Wang ZX, Teh CH-L, Kueh JLL, Lufkin T, Robson P, Stanton LW. Oct4 and Sox2 directly regulate expression of another pluripotency transcription factor, Zfp206, in embryonic stem cells. J Biol Chem 2007; 282: 12822-12830.

[49]   Kim J, Chu J, Shen X, Wang J, Orkin SH. An extended transcriptional network for pluripotency of embryonic stem cells. Cell 2008; 132: 1049-1061.

[50]   Takahashi K, Yamanaka S. Induction of pluripotent stem cells from mouse embryonic and adult fibroblast cultures by defined factors. Cell 2006; 126: 663-676.

[51]   Okita K, Ichisaka T, Yamanaka S. Generation of germline-competent induced pluripotent stem cells. Nature 2007; 448: 313-317.

[52]   Wernig M, Meissner A, Foreman R, *et al., In vitro* reprogramming of fibroblasts into a pluripotent ES-cell-like state. Nature 2007; 448: 318-324.

[53]   Wei CL, Miura T, Robson P, *et al.,* Transcriptome profiling of human and murine ESCs identifies divergent paths required to maintain the stem cell state. Stem Cells 2005; 23: 166-185.

[54]   Porter CJ, Palidwor GA, Sandie R, *et al.,* StemBase: a resource for the analysis of stem cell gene expression data. Methods Mol Biol 2007; 407: 137-148]

INDEX